X-RAYS:

X-RAYS:
MORE HARM THAN GOOD?

How you can protect yourself from unnecessary radiation
by understanding the uses and misuses of diagnostic x rays

by Priscilla W. Laws, Ph.D.

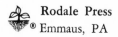 **Rodale Press**
® Emmaus, PA

2 4 6 8 10 9 7 5 3 1

Printed on recycled paper

Book design by Mike Anderson

Illustrations by Sidney Quinn

Library of Congress Cataloging in Publication Data
Laws, Priscilla W.
    X-rays: more harm than good?

    Includes index.
    1. Diagnosis, Radioscopic—Complications and
sequelae.   2. X-rays—Physiological effect.   3. X-rays
—Safety measures.   4. Consumer protection.   I. Title.
RC78.L363   616.07'572   76-58044
ISBN 0-87857-164-5

CONTENTS

A NOTE TO READERS

This book is primarily about x rays used for diagnostic purposes. Diagnostic x rays help physicians and dentists locate conditions in the human body which can cause illness.

X rays are also used for therapeutic purposes to kill cancerous cells, to destroy ring worm and other fungi, and to treat acne. Nuclear scans are sometimes used in which radioactive substances that give off x rays and other particles are introduced into the body for diagnostic purposes. Although these uses of radiation in health care also involve expense and exposure, they are not nearly as widespread as the use of diagnostic x rays. Therapeutic x rays and nuclear scans may, I hope, be the subject of other writings for consumers.

PREFACE

Since receiving my doctorate in nuclear physics in 1965, I have been teaching physics at Dickinson College. In 1972 I was granted a sabbatical leave to learn more about the environmental and health hazards of the low level radiation emitted from nuclear power plants. I started my sabbatical project by reading about the effects of radiation on human health and discovered that the diagnostic x rays used so commonly by physicians and dentists constituted a larger source of human exposure to man-made radiation in the United States than nuclear power plants, television sets, or high altitude jet travel.

I began to wonder if it was proper for my family dentist to be exposing me to x rays every six months. During my next checkup, I asked the hygienist who cleaned my teeth not to take x rays until after I talked with the dentist.

A few minutes later the dentist arrived looking tense, and asked why I had refused my x-ray examination. I explained that I had been reading about the health effects of x rays and was curious about the dose I would receive. He became furious and said that if I wanted to use the information to decide whether or not I ought to have x rays, I could find myself another dentist. He said he was using x-ray procedures developed at the best dental school in the country, that the state inspected his machine regularly, and that I had no right to question his professional judgment. I left his office in a state of shock, but the more I thought about our conversation, the angrier I became. Whether or not my dentist should conduct x-ray examinations, aside from the x-ray issue, I felt intimidated by his manner. Did he have a right to react with hostility to my question? I called the department of radiological health in my state. I found out from the health department that my dentist's machine was about average in terms of the skin exposure I was receiving. Next, I called the Pennsylvania Dental Association and asked about standard practice with respect to x rays during regular checkups. A representative explained that there was no standard practice for routine dental x rays, but that many dentists took full mouth x rays every five to ten years and then only took the less extensive "bitewing" x rays during checkups if there was some visual indication of a problem. Twice each year a hygienist had been performing bitewing examinations on me, my husband, and two children before the dentist examined our teeth. We all brush regularly, eat few sweets, and have few dental problems. This convinced me that if nothing else, my family was being subjected to unnecessary fees for all these x rays.

I then composed a letter detailing my experiences and sent copies to anyone I thought might be interested. The letter I sent to Ralph Nader produced a reply from Sidney Wolfe, a physician who is director of Nader's Public Citizen Health Research Group. The Health Research Group was already concerned about what it considered a growing problem of the overuse of medical and dental x rays. Dr. Wolfe invited me to come to Washington to discuss ways that I might help his group tackle the problem.

During my visit to HRG several staff members suggested that I write a consumer's guide to diagnostic x rays. Impressed with the fact that diagnostic x rays were already a larger source of public radiation exposure than nuclear power plant emissions, I dropped my project on nuclear radiation and spent the remainder of my sabbatical year and part of the next year researching and writing a consumer's guide.

Shortly after I began work on the guide, I met a local dentist at a social gathering who had heard secondhand about my experiences. We had a very candid and fascinating discussion about a range of topics including my experiences, dental x rays, his profession and its practices, preventive dentistry, and nutrition. He knew that I was a physicist and invited me to examine his x-ray equipment and make suggestions about how he could improve his techniques. He commented that he didn't believe in routine bite wing examinations every six months! Needless to say, my family was immediately established with a new dentist.

I began to realize how haphazard my family had been about choosing physicians and dentists. We had assumed that most up-to-date practitioners used similar procedures. An official at the American College of Radiology* who helped me gather information for the guide said that he and his wife now made a practice of interviewing physicians about their philosophy, practices, and fees before depending on one for care. With this in mind, I at-

---

* The ACR is a professional organization representing a large majority of physicians who specialize in the use of x rays and other forms of radiation in medicine.

tempted to include advice about choosing physicians and dentists and about what types of questions to ask about their x-ray practices in my guide.

Additional reading, discussion with physicians, and visits to the United States Bureau of Radiological Health convinced me that the question of how safe or dangerous x rays are is a complex one which is subject to considerable debate, and is not well understood even by physicians and dentists. I became increasingly aware of the fact that the radiation exposure from different types of x-ray examinations varies tremendously. I spent an entire summer making risk estimates on various types of x rays and found that some examinations, like the dental bitewing, involve very little radiation exposure while others, like abdominal x rays, involve much more. Since then I have been collaborating with another scientist at the Bureau of Radiological Health to refine these estimates and publish them. Thus, the original guide included an estimate of some typical risks associated with certain diagnostic x-ray procedures.

My efforts finally resulted in a guide to avoiding unnecessary x-ray exposure which was published in 1974 by Health Research Group, and is being distributed by them on a mail-order basis. The guide was designed to help intelligent consumers with no special training in science to learn more about what x rays are, how they behave, which examinations involve the greatest exposure, and how they are used and misused in medicine and dentistry. Most important, the guide was designed to arm consumers with intelligent questions they could ask physicians, dentists, and x-ray technicians in order to evaluate the safety of their radiation practices.

Consumer response to the original guide and the publicity surrounding it far exceeded my initial expectations. I received comments on my guide from physicians, dentists, nurses, radiologists, chiropractors and x-ray technicians. But mostly I heard from consumers. From all over the United States, hundreds of letters poured in filled with x-ray experiences, questions, and requests for advice.

Some of the letter writers who expressed concern to their physicians or dentists about the hazards of radiation said they were treated with ridicule. Others found that they were shuffled from specialist to specialist and subjected to the same examination many times without once being asked if previous films were available. There were people who had so many expensive and apparently useless examinations that they filed bankruptcy claims. Others were concerned about mistakes that technicians made which required repeat examinations, or the failure of technicians to shield their reproductive organs from x rays.

X Rays: More Harm Than Good?

In analyzing the correspondence, I began to see the problems caused by diagnostic x rays as only part of a much larger set of problems in the present health care system. It became apparent that mounting health care costs, malpractice suits, rising patient dissatisfaction with their health care, an increasing dependence on complex technologies, and the overuse of x rays are all related. Diagnostic x rays must be viewed in the context of these broader problems. For example, the costs of medical and dental care as well as the nature and perceived quality of care influence doctor/patient relationships. These relationships appear to have more influence on the likelihood of malpractice litigation than does the actual outcome of the patient's treatment. Many physicians and dentists freely admit that they are ordering more x rays than ever, not for their patients' benefit, but as a defense in case of malpractice suits. With a large proportion of the population currently covered by Medicare or by private health insurance, these medically unnecessary x rays appear to both the physician and patient to be "free." In reality this indirect method of paying for x-ray examinations is very costly. Current health insurance practices encourage the overuse of diagnostic x-ray examinations. This in turn spurs the use of more sophisticated x-ray devices and procedures for more and more medical conditions. The use of these complex technologies then raises our expectations about medical care, depersonalizes and fragments it, and makes it more expensive. These factors lead to further breakdown in doctor/patient relationships, adding to the malpractice problem, and so on in an ever-expanding upward cost spiral.

In this book I have attempted to supplement the practical information about x rays contained in the consumer's guide with a more detailed analysis of the situation, so that the problems associated with the use of diagnostic x rays are considered from the perspective of patients and health workers in the wider context of our modern health crisis. Each letter I received from a consumer reflects a deeply personal experience with x rays. Yet many common threads run through these letters. The personal accounts of those who have had unhappy experiences with x rays do not prove in themselves that the x rays are more harmful than beneficial. These experiences are not scientific studies. However, they do indicate that many people are dissatisfied with the costs and apparent ineffectiveness of x-ray examinations. The common problems of x-ray users described in this book may help other readers avoid certain pitfalls and understand more about health care practices as they relate to x-ray use.

I am not a physician or dentist. It would be presumptuous *and dangerous* for me to try to assess an individual's symptoms and make judgments about whether x rays are necessary or not. Although I have tried in this book to help you learn more about currently accepted x-ray practices, it will remain your responsibility to take part in deciding about when, where, and how you will receive x-ray examinations. It will be your responsibility to choose carefully those physicians and dentists from whom you will accept advice about x rays *and* other matters. This book is written with the conviction that you as an informed consumer can and should play an active role in improving the quality of your health care.

This book would not have been possible without the inspiration provided by Dr. Sidney Wolfe, M.D. and his fine staff at Public Citizen Health Research Group in Washington, D.C. It was the Health Research Group's profound belief in the power of informed consumers to affect change that motivated the original consumer's guide upon which this book is based.

I am grateful to Mark Barnett, MPH, Joseph Arcarese, MS, James Benson, MS, and Jim Morrison from the Bureau of Radiological Health, Otha Linton from the Washington Office of the American College of Radiology, and Robert Gorson, MS, from Jefferson Medical College for providing me with basic information about diagnostic x rays and reviewing parts of this book or the original consumer's guide. Radiologists John Harris, Jr., Theodore Bledsoe, and Albert Herwit and my present family dentist G. Ronald Krajack provided invaluable comments about the practical side of conducting diagnostic x-ray examinations. Although these professionals did not always agree with my conclusions about diagnostic x-ray use, I appreciated their input.

Several individuals provided editorial help and laymen's reactions to parts of the manuscript. These individuals include Wendy Cheyfitz, Marcia Conner, and Frances Watson.

Last but not least, I would like to thank the several hundred "patients" throughout the United States who wrote me about their actual diagnostic x-ray experiences.

Priscilla W. Laws
Carlisle, Pa.
September 1976

CHAPTER 1
X RAYS—USES, BENEFITS, COSTS, AND RISKS

Medical advances beginning in the 1930s and extending through the late 1950s brought about significant improvements in health, especially through the control of infectious diseases . . . . For more than a decade, however, the impact of new medical discoveries on overall mortality has been slight; indeed, the death rate for U.S. males at most ages . . . has actually been rising.

Victor Fuchs
Who Shall Live?

The fact is, the United States pays more per capita for health care than any other industrialized nation in the world, but it gets less health care.
Senator Edward Kennedy
In Critical Condition

Introduction

In spite of the fact that the number of diagnostic tests, medical treatments, and drugs used by the average individual is greater than ever, mortality rates for all ages are increasing in the United States. Critics of modern medicine claim that the longer life spans and decreases in infant mortality achieved earlier in the twentieth century were due primarily to environmental factors related to improvements in housing, sanitation, diet, and public health, and to certain very specific medical advances such as the development of vaccines and antibiotics with which many communicable diseases can now be prevented or treated. The more recent increases in mortality, especially from cancer, are being attributed in part to the presence of more cancer-producing chemicals and radiation in our physical environment. Personal habits like eating high-cholesterol foods and smoking have contributed to increases in mortality from heart disease, lung cancer and emphysema.

One can only conclude that the nation's health has at least as much to do with environmental factors and individual habits as with medical care. In fact it may be the case that modern medicine has a negative influence on national health. Ivan Illich, author of *Medical Nemesis,*[1] claims that an expanding proportion of disease is medically caused, or iatrogenic. Examples of medically caused diseases include the side effects of modern drugs, damage to children whose mothers received diagnostic x rays during pregnancy, and the pain, anxiety, and expense endured by unsuspecting patients who become needlessly involved with extensive medical tests.*

Many consumers wrote to me about the extreme discomfort, distress, and anxiety they suffered when subjected to extensive x-ray examinations. Although the potential risks from radiation are not immediately detectable in most cases, the illness-producing or iatrogenic side effects of their experiences are clear. For example, a 63-year-old man who read about my consumer's x-ray guide in the newspaper wrote to me. In 1972 he visited his ophthalmologist for a regular checkup. The ophthalmologist claimed that the patient had abnormal pressure in his left eye. After consultation with the family physician and a neurologist, the patient was informed that unless he submitted to a series of x-ray examinations and subsequent treatment, he would be an excellent candidate for a heart attack due to arterial blockage in the left side of his head. The patient described his x-ray experience as follows:

> I had 59 shots taken of my head, lungs, chest, and spine in one morning and a total of 87 in the one day. (One of the technicians admitted later that 33 of the 87 x rays were "defective.") A few

---

* Illich cites evidence in *Medical Nemesis* that most of today's skyrocketing medical expenditures result from diagnoses and treatments of doubtful effectiveness.

hours after all these I collapsed while walking in the hospital corridor. I was perspiring profusely and extremely weak, vomiting, and had a very rapid pulse rate. The hospital was alerted and responded, but all that was done was to pick me up and place me in bed. . . .

The tests cost the insurance company $1,500 and perhaps me in the near future my life. The tests showed absolutely nothing wrong anywhere.

Since this, I have had other eye tests and of three good eye doctors none has found low eye pressure. Still, since all of these x rays and those of years past I seem to be getting weaker, with continual pain in my muscles. I still cannot swallow food without partially choking because of the extreme pressure which was applied to my throat area for the carotid artery x-ray angiography.

Something *is* wrong with me just since these x rays and tests but I do not know what it is except that I was fine and *very healthy* before the x rays and tests.

As a result of reading my consumer's x-ray guide, a man wrote to me on behalf of his disabled mother who was badly injured in an industrial accident in 1960. Her employer's insurance company refused to pay for the treatment she needed and she consequently filed an industrial accident claim. The attorney appointed to represent her referred her to a physician who ordered a special x-ray examination of her spinal cord. This exam, known as a myelogram, usually requires the injection of a special dye into the spinal cord. The injection of the dye can be risky. The woman's son described her experience:

She never signed a consent form to have the myelogram performed. During the adminstration of the myelogram her right leg began to jerk very badly, and when the anesthetic wore off, the bottoms of both feet started drawing up. She had spasms in both feet. She was very sick to her stomach, and her eyes and head began to hurt.

Shortly after the examination she couldn't eat or stand or walk without the aid of crutches.

A woman who read about my x-ray guide in a health magazine wrote to me about her experience. She had extensive skull x rays in 1970 to diagnose the reasons for fainting and blind spells. Three years later she developed a "frightening-looking spot—crusty and red" on the top of her head and consulted with a dermatologist who gave her some ointments. Two years later she was still troubled by the spot and she consulted with an internist and another dermatologist. The dermatologist "insisted that the spot was not due to x rays." She wrote:

> My son, who lives in another town, consulted a dermatologist with whom he is acquainted and told him about my circumstances. He told my son that I probably have an x-ray burn but that no doctor in my town is going to admit it because they all work at the same hospital. The dermatologist predicted that I will have problems with my scalp the rest of my life.

In addition to such individual reports of ill effects which patients believe were caused by improper use of x-rays, there is scientific evidence that diagnostic x rays of unborn children and of the abdominal area in adults may increase the incidence of leukemia ten to fifteen years after exposure.[2] X rays can damage the reproductive cells of potential parents, subsequently causing genetic diseases in future generations.[3] Thus some diagnostic x rays can indirectly cause leukemia. The incidence of x-ray-produced genetic diseases or leukemias are examples of iatrogenic disease.

Although most physicians believe that at present the overall benefits of diagnostic x-ray examinations probably outweigh the risks, an increasing number of professionals feel that diagnostic x rays are often misused and overused by medical and dental practitioners.[4] For example, a report published recently by the Environmental Protection Agency states that ". . . it appears reasonable that as much as a 50 percent reduction in dose might be possible due to [improvement in] technical and educational methods."[5] Regardless of whether or not this estimate is accurate, many qualified professionals feel that a significant portion of the present annual population exposure to diagnostic x rays is unnecessary.[6] Thus, even though x-ray studies have been of great benefit to many people, their potential for harm should be considered before a decision to conduct an x-ray examination is made.

How often do people use x rays? What do they cost? What types of benefits and risks can we expect to encounter when x-ray examinations are prescribed?

The Costs and Extent of Diagnostic X-ray Use

The number of annual x-ray examinations has been going up dramatically in the past fifteen years. In 1964 an estimated 77 out of every 100 people visited x-ray facilities. By 1970 Americans were making an estimated 179 million visits to x-ray facilities each year at a rate of 90 visits per 100 people.[7] The x rays cost approximately 3 billion dollars and represented about 4 percent of the total national expenditure for health care in that year.[8]

Because of rising costs and frequency of x-ray examinations the total costs of diagnostic x rays were expected to rise to 4.8 billion dollars by 1975.[9] This means that the average family of four in the

TABLE I:

ESTIMATED NUMBER AND COST OF DIAGNOSTIC EXAMINATIONS IN 1970 *

Examination Type	Number of Exams (Millions)	Average Cost Per Exam	National Expenditure (Millions)
I. Medical X rays	142	$17.19	$2,440
II. Dental Examinations: 1. Bitewing	27.2	6.85	186.5
2. Bitewing and Periapical	8.9	8.50	76.5
3. Full Mouth Series	7.4	17.67	130.7
4. Periapical	19.9	5.00	99.5
5. Other (Panoramic Cephalometric, etc.)	4.4	17.67	77.7
6. Total Dental	67.8		570.0
III. TOTAL MEDICAL AND DENTAL	215		$3,010

* Based on Data from the U.S. DHEW Environmental Assessment Report on Performance Standards for Diagnostic X-ray Systems and Their Major Components (August 1974).

United States may now be paying more than $85 each year for medical and dental x rays. Chiropractors also use diagnostic x rays. Some conduct relatively high-dose whole body and spinal examinations on patients, and since there is one chiropractor for every ten physicians, additional millions of dollars are probably being spent annually for chiropractic x rays. In addition, x-ray procedures are being used in penal institutions to search for concealed drugs or weapons. The number of such exposures is unknown.

The Benefits of Diagnostic X rays

Historically, diagnostic x rays have been used for the detection of metal objects, cavities, and broken bones. Films of various parts of the body such as the chest, skull, abdomen, and extremities have been taken ever since x rays were first discovered in the late 1800s. More recently, techniques have been developed in which a foreign substance such as air or a chemical which absorbs x rays easily is introduced into the body to "contrast" with surrounding organs and tissues. Contrast procedures are now commonly used to study the digestive system, kidneys, urinary tract, arteries, blood vessels, and heart valves. New techniques in fluoroscopy allow physicians to study the motion of body organs and tissues on television tubes.

The battery of available x-ray examinations allows physicians to detect abnormal growths or lesions in internal organs which can lead to such common causes of death and disability as breast cancer, lung cancer, and cancers in the gastrointestinal system. Conditions such as emphysema, heart disease, infections of the kidneys and urinary tract, ulcers, and pneumonia can be revealed, as well as the existence of tumors, cysts, and polyps. Chronic illnesses which x rays can detect include arthritis, tuberculosis, and bone demineralization.

Dentists, periodontists, orthodontists, and oral surgeons can use x rays to check for jaw fractures, tooth alignment, decay between teeth, gum diseases, abcesses, impacted teeth, hidden tartar deposits, and bone cancers.

It is accepted that diagnostic x rays can be very beneficial. Setting broken bones and removing unwanted tumors would be difficult, if not impossible, without x rays. However, modern diagnostic x-ray procedures are sophisticated, time consuming, and expensive. Certain procedures involve risks due to the use of the contrast materials and the absorption of x rays. Thus, x rays should never be used casually.

Facts and Fallacies about Risks

The relationship between exposure to different types of radiation and health is very complex. I have read and heard a great deal from

6

both professionals and lay people about radiation risks that is just plain false. On the one hand, it is easy to carry a fear of x rays to dangerous extremes. For example, one woman asked me how she could tell if she had too many x rays. "I lie awake nights worrying about getting cancer," she said. Another wrote that "I fear for my life." Some people attribute health problems such as bad nerves, stomach cramps, poor eyesight, and back pains to "radiation overdose." In a similar vein, Griffiths and Ballantine, extreme critics of diagnostic x-ray practices, wrote in their book *Silent Slaughter* that "X-ray machines . . . are killing you with low-level radiation. Cancer, leukemia, congenital deformities are likely consequences. . . ." [10] This would certainly appear to overstate the actual risk.

It is easy to carry a fear of x rays to dangerous extremes. For example, a number of well-educated women in the New York area indicated during a telephone survey by the New York Public Interest Research Group that they would not want to submit to an x-ray examination for breast cancer even if they had lumps. For women with lumps, x rays can serve as an invaluable aid in the detection and treatment of breast cancer. It is also possible to over-react to the apparent danger of fetal x rays. *Good Housekeeping* magazine, in a story entitled "A Nightmare Decision," described an experience of a young mother.[11] A genetic counselor had advised this woman to have a therapeutic abortion because she had one abdominal x-ray examination early in her pregnancy. On the basis of the information provided in the article, the risk of fetal damage seemed small compared to the emotional and physical trauma the abortion would cause to the mother.

At the other extreme, those who prescribe and conduct examinations tend to underplay the risks. A young x-ray technician (radiologic technologist, or R.T.) who received her training in a two-year hospital school explained in a letter that "hospital students are taught by staff technicians, and few of them are knowledgeable about radiation dosage or dangers." She reports that during her six years as a technician she picked up a number of standard answers for patients who question radiation dosage, such as "the dose is minimal," "you're receiving very little," "it's so small it won't make any difference," or "it's unlikely [that] you will receive enough x rays in your lifetime to do you any harm." She goes on to say that pregnant women receiving abdominal x rays are rarely informed of the added dangers to the fetus.

A dentist wrote, "I have taken tens of thousands of x rays without adverse harm to any of my patients or to myself." A number of dentists, physicians, and chiropractors tell patients that their x-ray examinations "involve less radiation than a stroll in the sun." Many physicians have claimed that there is no proof that the low levels of radiation involved in diagnostic x rays have ever harmed anybody. One doctor commented to me in a letter that "a person would have to have 100 x rays before there could be any danger."

After searching scientific literature on the subject and giving the matter a great deal of thought, I am convinced that the truth about radiation effects lies between the two extreme views on risk. There are elements of truth and falsehood in all of the claims.

For example, the ultraviolet radiation from the sun and the x rays generated in a machine have similar properties, but medical x rays are more energetic—more penetrating. The amount of radiation dose to the skin and resultant risk of skin cancer can be similar for both a stroll in the sun and a diagnostic x-ray examination. However, diagnostic x rays can penetrate and be absorbed by other organs and tissues in the body. Some of these organs, such as the red bone marrow, the thyroid, and the breasts can develop cancers or leukemia from five to 30 years after exposure to radiation, although there may be no immediately detectable effects of x-ray doses.

There are studies which have linked abdominal and lower back x rays of adult males with an increased chance of eventually developing leukemia. There are also several studies which indicate that diagnostic x rays of pregnant women which place unborn children in the main beam cause an increased chance of childhood leukemia. In general children are more susceptible to radiation damage than adults.

It is reassuring that the risk of developing cancer or leukemia later is not large for more procedures. For the higher dose procedures such as those involving annual x rays of the lower back or abdominal area, the risk of developing radiation-induced illness is probably similar to the risk of lung disease incurred by a regular cigarette smoker. For lower dose procedures like dental and chest x rays, the risk factors are considerably smaller.

There is no direct evidence that diagnostic x rays of other regions besides the trunk are harmful to adults. However, many scientists believe that x rays may cause damage in proportion to the amount of dose involved, and if the reproductive organs are in the direct x-ray beam, such x rays may potentially damage future generations as well as exposed individuals. If risks are proportional to dose, then there is no safe level. Every examination adds its share to the cumulative risk, and it makes no sense to say a person "has not had too many x rays yet."

There are additional risks associated with the x-ray examinations which include the injections of contrast media to outline certain organs, arteries, or blood vessels. In many instances, these risks are minimal and consist of the possibility of mild allergic reaction to the contrast medium. However, some contrast studies, like those of the heart or the brain, can cause strokes or other serious complications if not performed with great care.

Unfortunately, physicians and dentists are not always as well informed about the health effects of x rays as they should be. In 1976 Baltimore City Health Commissioner John DeHoff stated:

> . . . the majority of practicing physicians who are unfamiliar with physics or epidemiology and who don't read research reports in physics journals have no visualization of what may happen when radiation passes through a body.

Physicians, dentists, or x-ray technicians who insist that x-ray examinations are completely harmless are either being intentionally deceptive or are misinformed themselves. High dose procedures should not be conducted routinely or without careful consideration on the part of the patient and prescribing physician.

CHAPTER 2
RELIEVING ANXIETY
WITH X RAYS

Relieving anxiety is a large part of almost every physician's stock-in-trade. This 'non-curing' role of the physician takes many strange forms.

Victor Fuchs
Who Shall Live?

Too often, the x ray to an adult is as the band-aid is to the child—it, in itself, makes him feel better!

A chiropractor, 1976

A person who is worried about his or her health may decide to see a doctor. The doctor may refer the patient to a radiologist for diagnostic x rays. This is normal procedure. However, it is not always the course of events. I once asked a private radiologist if he was willing to deviate from normal practice and directly x-ray an individual requesting an examination. He immediately said yes, and pointed out that such people are usually quite worried about their health. The radiologist was convinced that the positive benefits of relieving anxiety in his patients far outweighed what he considered the minor expense and risks involved. He then told me about a heavy smoker who had recently come to his office. The man was deathly afraid of lung cancer. "I took several chest x rays and told the man nothing was wrong with his lungs. He looked extremely relieved, lit up a cigarette, and thanked me profusely as he left the office."

11

Cultural Attitudes Towards Medicine

The story about the radiologist and the heavy smoker says something significant about our attitudes towards medicine. Consumers and the health care establishment are heavily geared towards the diagnosis and treatment of illnesses rather than the maintenance of health. This emphasis is not always the fault of the health establishment. In many ways we get what we ask for. My current family dentist, who is a leading proponent of preventive dentistry, told me about a young woman who came to him with a mouthful of cavities. He took x rays, located and filled the cavities, and then explained the importance of proper diet and of regular flossing and brushing to prevent further problems. Six months later she returned with more cavities. When the dentist asked her why she hadn't taken care of her mouth, she blurted out, "Oh, Doctor, it's too much trouble to do all those things. Why don't I just come in regularly and let you fix my teeth?"

Our modern lifestyles cause us to do many things which help destroy our health. We overwork, suffer from stress, fail to exercise, and are exposed to environmental pollutants. But above all we are a nation of consumers. We consume alcohol, tobacco, junk foods, and prescription drugs in tremendous quantity. Critics of the health care system are beginning to recognize that personal habits and lifestyles have more influence on health than all the diagnoses, treatments, and cures of the medical system. We have a habit of exposing ourselves to health hazards, and then we depend on modern medicine to provide a technological fix—to make things right again.

A number of factors contribute to our present attitudes towards medicine. For one thing, we have a natural fear of pain and death. But Ivan Illich feels that because modern medicine can kill pain we have lost the incentive to avoid it by managing our own health. The success of medicine in lowering infant mortality and preventing diseases like smallpox and polio has led us to believe that early death and disability can be prevented. The miracle of heart transplants has convinced us that surgical procedures can at least postpone death.

The belief in technological solutions to all our problems is widespread in America. We believe that population control can be accomplished with contraceptives, solid waste problems can be solved with

12

expensive modern treatment plants, the development of nuclear fusion will solve our energy problems, and modern medicine will eventually eliminate pain and death.

Our faith in a highly sophisticated and technologically oriented health care system has led us to an overdependence on physicians and dentists, and we tend to blame them when things go wrong. Unfortunately, the medical profession has done little to discourage the public image of its omnipotence. Medical professionals are frequently unwilling to admit that there are any conditions they cannot diagnose or treat. If all else fails the condition can be labeled as psychological in origin and treated with tranquilizing drugs. Maintaining the myth of omnipotence contributes to the problem of unnecessary x rays in at least two ways. First, the unwillingness to admit that there may be no diagnosis or treatment for certain patients leads physicians to order more and more expensive and elaborate tests for these patients. Second, the profession is reluctant to admit that any of its members are incompetent. This lack of self-regulation contributes to the growing incidence of both malpractice and malpractice litigation (some of which may not be a result of true malpractice). In turn, the growing malpractice problem creates a climate in which physicians practice defensive medicine and order unnecessary x rays and other expensive diagnostic tests to protect themselves in case of lawsuits.

According to Illich, our faith in modern medicine results in a loss of what Illich calls our "potential to deal with human weakness, vulnerability, and uniqueness in a personal and autonomous way." Illich claims that this is a type of "cultural iatrogenesis" which "occurs when people accept health management designed on the engineering model, when they conspire in an attempt to produce, as if it were a commodity, something called 'better health.' " [1]

A young woman who recently had a series of diagnostic examinations wrote to me after reading an article about my x-ray guide. She underwent a change in attitude towards medical care which beautifully summarizes our responses to anxiety as well as the pitfalls of expecting medical solutions to all our problems. She writes:

> I . . . acquired an abdominal pain in the middle right side just below the rib cage. I dealt with the pain for about ten days and

13

then sought out a general practitioner.

At first he gave me laxatives and pain pills . . . and said that if I didn't feel better in three days come back.

I came back and he did an urinalysis which came out negative so he scheduled me for x rays.

First, I was x-rayed for gallbladder problems, then kidney, then upper GI and finally lower GI—all were negative.

The doctor decided it was psychosomatic gastroenteritis. (I found this out by accident when I got a letter from Medi-Cal, which I wasn't eligible for, with the doctor's diagnosis and fees.)

I called the doctor again and said I didn't feel any better and that I had diarrhea and a bloated abdomen. He said that he would schedule some more x rays.

I discontinued seeing him and have no faith in the present philosophy of the medical profession. I still have the problem but have reduced its magnitude measurably by good diet and exercise.

This patient has clearly shifted the responsibility for her care from the physician to herself.

X rays as Placebos

"Placebo" is a term applied to any harmless medication such as a sugar pill which is given to a patient under the pretense that it is an active drug. The interesting thing about placebos is that they appear in many cases to be effective. Patients who believe in the potency of medicine feel better after taking placebos. Jerome Frank, a psychiatrist who has studied placebo-use notes that the belief in some treatment can actually enable patients to overcome the effects of certain drugs.[2] Frank describes an experiment in which patients were given a drug intended to cause stomach contractions and vomiting. When told that their stomachs would not become upset the subjects overcame the effects of the drug—an apparent case of mind over matter.

The reasons for the effectiveness of placebos are not well understood, but Dr. Frank has proposed that the placebo:

. . . gains its potency through being a tangible symbol of the physician's role as a healer. In our society the physician validates his power by prescribing medication.[3]

It is obvious that prescribing x-ray examinations solely for their psychological impact is like ordering a placebo for an anxious patient whose complaint cannot be diagnosed. The ability to order x rays and tests seems to validate the physician's power as a healer, and the x ray symbolizes for us tangible proof that something either is or is not wrong with our bodies. Even though many medical conditions cannot be detected by x rays, a negative x-ray examination serves the same psychological function as a sugar pill disguised as a "potent drug." Both assure an anxious patient that he is going to be all right, and this belief seems to improve the individual's health.

Dr. David Glasser, Baltimore Chief of Communicable Diseases, commented in a 1976 newspaper interview, "Some people—it's amazing—they actually feel better after getting irradiated—like it's some sort of therapy."

If they are effective, what's wrong with using x-ray examinations as placebos? The use of x rays as placebos differs from the prescription of harmless medications in two ways. First, x rays are not harmless and second, they are more expensive than most drugs. (Other objections to x rays as placebos are similar to those leveled against the use of drugs as placebos.)[4] It has been noted that one reason physicians resort to prescribing placebos so readily is that it allows them to write a prescription and get on to the next patient without wasting any time. How much easier to write a tranquilizer prescription for an anxious patient than to expend the time and effort needed for consulting. The pen is flourished, the prescription form is ripped off the pad, and a drug (or x-ray examination) has been ordered. The danger of this approach is that it can be used in place of an adequate physical examination or as a substitute for real efforts to deal with a serious problem. Furthermore, the acceptance of x rays as placebos in the profession legitimizes a form of deception which if extended to other areas of medical care seems unwholesome and counter-productive.

If a physician genuinely feels that an x-ray examination or prescription has no value except for psychological purposes, it would seem more honest to take the time to explain the situation carefully to a patient. This affords the patient dignity and saves him money. The physician must also learn how to talk to patients about their problems and serve as a person who cares rather than as a conduit in a complex system of tests and therapies.

In his book entitled *The End of Medicine,* Rick Carlson expresses the feeling that placebos are needlessly expensive, and that equally effective but less expensive symbols of healing must be sought.[5] More and more people are turning away from traditional medical approaches to their health problems. The upsurge in popularity of the Adelle Davis books and publications like *Prevention* magazine, which encourage people to maintain their own health with good diet and vitamins is testimony to this. In addition, more and more people are beginning to use diet, exercise, and vitamins to help treat their own symptoms of disease and ill health. It is impossible to estimate how much the "placebo effect" has contributed to the success of this approach. However, people who are able to treat themselves successfully in this manner are gaining a healthy sense of self-respect and autonomy. They are finding these self-help methods much cheaper than medical care.

Although the programs of self-administered health care involving diet, vitamins, and exercise are cheaper and less prone to iatrogenic side effects than many medical tests and treatments, I don't mean to imply that we can ignore the contributions of modern medicine. But it is important to realize that this self-regulated approach to health care will be more beneficial if we have a deeper understanding of what medicine can do and what it cannot do.

Illnesses can be divided into three categories—those which are self-limiting, those which can be prevented or cured with proper medical treatment, and those which, at present, cannot. Self-limiting diseases such as the flu and common cold run their course naturally. Given rest and proper nutrition, the body has an amazing capacity to heal itself. Yet countless people fearing something more serious indulge prematurely in costly tests, rather than placing any trust in the body's ability to heal itself.

Modern medicine has, of course, proven its effectiveness in treating certain conditions. Antibiotics and sulfa drugs have been extremely useful, although they tend to be overused, in eliminating a number of bacterial illnesses. Infectious diseases such as diphtheria, tetanus, and polio can now be successfully prevented. Some illnesses, however, resist our best efforts to effect a cure. It is commonly accepted that modern medicine cannot cure heart disease and stroke once they have occurred. Although cervical cancer and some types of skin cancer can be effectively treated, it appears that most cancers cannot be cured. Some of the most tragic of all the conditions which cannot be cured are chronic ailments like arthritis and multiple sclerosis. Victor Fuchs observes that the medical treatments which are most dramatically effective, such as vaccines and anti-infective drugs, are cheap and relatively easy to administer.[6] Those treatments which attempt to deal with presently incurable conditions tend to become progressively more technical, sophisticated, and costly.

Consumers must somehow learn more about the strengths and weaknesses of modern medical care. We must learn to use restraint when a condition appears to be self-limiting, and learn how to care for ourselves. We must learn to face the possibility of pain and death with dignity in cases where there is little hope of effective diagnosis or treatment—avoiding the personal and financial bankruptcies which so often accompany serious illness. We must know when it is best to go and take advantage of what modern medicine has to offer us.

Learning more about diagnostic x rays is just part of the much larger task of taking active responsibility for our health care. Expecting too much or too little from x-ray examinations and other tests and treatments can only lead to needless anxiety and hostility towards the medical establishment.

CHAPTER 3
THIRD PARTY PAYMENT

Hospitals register well-insured patients, and rather than providing old products more efficiently and cheaply, are economically motivated to move towards new and increasingly expensive ways of doing things . . . increased insurance coverage . . . encourages hospitals to provide products more expensive than those the customer actually wants, needs, or would have been willing to pay for directly.

Ivan Illich
Medical Nemesis

There is little doubt that the overuse of diagnostic x rays is encouraged by the existence of insurance programs. Insurance coverage makes most of us more accepting of extra x-ray examinations and tests. After all, we are not paying for them directly. In this chapter we will explore some relationships between insurance and diagnostic tests, including x-ray examinations.

The cost of personal health care now averages over $250 per person each year and continues to increase. The amount of medical care needed by an individual is highly variable from year to year. People who suddenly become seriously ill can unexpectedly face medical bills totaling tens of thousands of dollars. Insurance is the most natural method for sharing the financial risks of ill health in an industrialized society.

The years between 1950 and 1975 have seen a rapid growth both in private and federal insurance programs. In 1950 third parties paid for 32 percent of the nation's personal health care, but by 1974 this had increased to 65 percent of the total.[1] This expansion of insurance coverage was accompanied by a rise in medical care costs which has been notably greater than the rise in prices for other goods and services. Although the increase in medical costs has been attributed to the introduction of more expensive technologies and the ever-increasing utilization of diagnostic tests and therapies, health insurance has been primarily responsible for the whole trend towards more extensive and expensive health care. In the mid-1960s, the federal Medicare and Medicaid programs were added to the already expanding private insurance companies. Selig Greenberg writes in his report on the condition of American medical care:

> Health-care costs took off on a new orbital course after the inauguration of the Medicare and Medicaid programs in mid-1966. . . . In a three-year period following the start of these programs, hospital charges ballooned by 55 percent and doctors' fees rose by more than 25 percent, or two-and-one half times their previous rate of increase. This has sharply raised federal and state expenditures for medical care and led to a rash of successive increases in Blue Cross premium rates.[2]

Physicians are much more casual about ordering elaborate diagnostic tests and treatments if there is insurance coverage for them, and hospitals are more apt to install the latest technological devices to conduct them. Patients seem to view these costly tests with pride. The patient does not pay directly and so they appear to be free. They also provide tangible evidence of the benefits gained from employers, Blue Cross, or Medicare. What is often forgotten by physicians and patients alike is that the public eventually pays for care under any system. There is no way to transfer the cost of health care to business or the government without individuals paying more taxes or higher prices or receiving lower wages. Different systems of health insurance can influence the proportions of care paid for by different income groups, but a redistribution of the burden usually only affects the high and low income families. The average family will probably pay the same share under any system.

Insurance does, however, have the effect of diverting resources from other sectors of the economy and thus increasing the overall costs. This way in which insurance as a method of risk-sharing influences costs is analogous to the tragedy of the commons described by the environmentalist Garrett Hardin.[3] The tragedy is roughly as follows: A group of herdsmen share a large meadow, or commons, for grazing their sheep. When each herdsman has one sheep, the commons is almost filled to capacity. One herdsman discovers that if he grazes one extra sheep on the commons he will be twice as wealthy and the commons will still be able to support all the sheep. One by one the other herdsmen realize that they have just as much right to use the commons as their neighbor with two sheep. As more and more herdsmen acquire extra sheep the commons becames overgrazed. The top soil erodes away and the capacity of the commons to support the sheep and hence their herdsmen is destroyed.

In our eagerness to obtain more than our "share" of the medical goods and services paid for by health insurance (the commons), we add to our own as well as to the overall financial burden which all of us must pay, as well as to the level of iatrogenic or medically produced illness.

Private Insurance

An estimated 80 percent of people under the age of 65 had private medical insurance coverage by 1973, and essentially all of those with private insurance had at least some coverage for diagnostic medical and dental x-ray examinations.[4] Private health insurance can be grouped into three categories: Blue Cross/Blue Shield, commercial insurance companies, and independent plans. In terms of the number of people covered and the total number of claim dollars paid, Blue Cross/Blue Shield and the commercial insurance companies share the bulk of the business about equally. Independent plans like the Health Insurance Plan in New York, the Kaiser Plan in California, and others, serve significantly fewer people. It is interesting to note, however, that patients in the independent plans, which are prepaid, are subjected to one-third fewer hospitalizations and significantly less expense than people enrolled in Blue Cross/Blue Shield. This has been attributed to the fact that many doctors are admitting patients to the hospital because the insurance companies will cover the costs of tests to hospital patients but not to outpatients. Once a patient is in the hospital it becomes easy to run a whole battery of extra tests. However, the independent plans usually provide a range of diagnostic tests without requiring hospitalization.

Hospitals and insurance companies may even collaborate without the knowledge or consent of patients in providing unwanted services. One angry individual sent me copies of correspondence with his commercial insurance underwriter:

> Recently my wife was a maternity patient at the Harrisburg Hospital. Two days prior to discharge she was subjected to chest x rays. A $12 charge was made by the hospital for the use of facilities and a separate $8 charge for the radiologist. Upon investigation, it appears this service was performed NOT at the direction of the attending physician NOR as the result of any symptom of disease or injury. The attending physician advised that this procedure was undertaken as the result of "hospital policy." Further investigation disclosed that this item is covered by Blue Shield, and in the case in point was paid by your insurance company. I strenuously object to this approach whereby the patient is not consulted or given a choice concerning a matter of this nature. I likewise feel that insurance carriers should take steps to prevent payment of similar bills, so incurred, which undoubtedly contribute to rising premium costs. The individual involved in this instance was a maternity patient neither desiring nor requiring a chest x ray. I realize the possibility of misinformation or misunderstanding but submit this for your observations.

The insurance adjuster replied:

> They (the hospital administrators) said it is hospital policy to take chest x rays of patients. The policy was formulated to assist the community in the early detection of respiratory ailments. They feel it has been quite successful. They agree that there are many times when the x ray produces a negative result. Hence, it appears to the patient that it was totally unnecessary. They go on to say, however, that when the x ray results in the detection of a respiratory infection that has not yet manifested itself in physical symptoms, the patient is immediately given early treatment with a much better chance of cure. The knowledge of the respiratory illness allows the patient to guard against further infection of family and friends.

They feel that the savings of economic loss caused by early detection (expensive treatment costs are many times avoided—decreased loss of work time—no further infections of other people—closing of many expensive hospitals used solely for respiratory ailments) far exceed the expense of this routine x ray. I hope this letter explains to you the routine x ray. They are quite sorry that the only reply you got was "it's hospital policy."

The aggrieved individual replied:

It is interesting to note the broad statements concerning the results of these procedures. Are they substantiated by fact? How many hospitals—how many patients—for how many years? I appreciate that it is impossible to place a dollar value on life or health, but I rebel against decisions such as this without any consent by the patient. The local paper this weekend carried a schedule of free x-ray examinations by the TB society, available to all for the same purpose advocated by the medical industry. The main difference was that it was a determination for the individual. I wish to reiterate that the $20 involved in this one minor case is not the question. It is the policy that permits the medical industry to conduct tests without permission of the patient, AND THEN IS subsidized by the insurance carriers. Why should hospitals make these decisions? If, as claimed, the results justify the means, it would stand to reason that electrocardiograms and Pap tests should become hospital policy. Certainly there can be no question that heart disease and cancer are greater problems than respiratory disease, and the tests involved do not include radiation, a procedure which at best is arbitrary. It appears that the consumer lacks the pressure of lobbyists or the will to resist. It is time the medical profession and the insurance carriers recognize that consumers also deserve a voice in policy, especially since they are one party so vitally concerned.

The final sentence above points to the need for effective consumer representation on the boards of directors of private health insurance companies. These boards usually are dominated by insurance company representatives, hospital administrators, physicians, and other health care personnel. Expensive policies like those leading to added hospitalization

23

and diagnostic x rays without patient consent will not be eliminated unless patients demand more voice in formulating these policies.

Government Insurance

By 1974 over half of all public spending for health care was accounted for by Medicare and Medicaid. Each program costs over 11 billion dollars a year and together they account for 25 percent of total personal health care expenditures.[5]

Medicaid, originally established in 1951, combines federal and state money to provide financial assistance for the poor. Because the program is administered by the states, it varies tremendously in comprehensiveness. It is administratively complex and there have been reports that the program has stimulated fee-gouging and other abuses on the part of doctors.

Medicare, which was established in 1967, presently provides partial coverage for hospitalization and outpatient care for people over 65 as well as disabled individuals and those suffering from chronic kidney diseases. Included in the outpatient services are diagnostic x rays, laboratory tests, and radiation therapy. However, for outpatient insurance the patient must pay a monthly premium, the first $60 of the cost of care, and 20 percent of any costs above this amount. There is little dispute about the fact that both Medicaid and Medicare encourage the overuse of diagnostic tests. Cities which have large retirement populations abound with new store-front clinics which perform diagnostic tests paid for by Medicare. An elderly man walking out of a store-front clinic commented to a television interviewer that "you go in with an aching ear and get x rays from head to toe." The fact that Medicare enables doctors and clinics to indulge in expensive diagnostic tests and yet does not pay full costs, leaves patients in a double bind. Recently an older woman who had injured her shoulder in a fall wrote to me. She received a chest x ray which showed a shadow on her lung. Her doctor then ordered "25 x rays, an electro-cardiogram, and several blood tests." He asked her to go into the hospital for more tests. She informed him that all she had was Medicare and asked him who was going to pay for the tests. He said, "Don't worry about it." Her stay in the hospital lasted a week and involved 75 x rays, brain scans, bone scans, two biopsies, a bronchoscopy, and several blood tests. Her doctor wanted to test her pancreas and if he found nothing he said he would open her chest.

> When I heard that I got dressed real fast and told him I wanted to leave the hospital. His face got scarlet and he said, "OK, but your blood is bad and I wouldn't be surprised if you have the same thing as Martha Mitchell." That was three weeks ago. I now feel fine. I spoke to my former doctor who is retired. He said that he saw the records and they don't show a thing. The bills for over $3,000 are still piling up. I'm moving at the end of the month and will file bankruptcy as I haven't the money to pay for them.

A seventy-year-old woman who was subjected to numerous x rays commented, "If he took one, he took a hundred. I wouldn't be surprised to read in the *New York Times* that doctors were legally padding Medicaid bills by giving unnecessary x rays."

Recently payment for chiropractic services has been allowed by the Medicare program. The effectiveness of chiropractic treatments is viewed with suspicion by the conventional establishment. The result of this suspicion has been to encourage more x rays by chiropractors. This is because Medicare requires chiropractors to x-ray their patients who seek Medicare coverage in order to be sure that they have an abnormality amenable to chiropractic treatment. This requirement exposes many patients to unnecessary radiation and personal expense in order to be eligible for Medicare reimbursement. One chiropractor who read my guide wrote:

> I share your concern about overutilization of diagnostic x rays. In my practice, I try to avoid unnecessary x rays. I find that by being very exacting in examination and history taking, x rays are not always necessary to make a diagnosis. . . . The Medicare regulations will pay for chiropractic care *only* if x rays are taken, and the x rays must be re-taken in a year's time. I feel this encourages unnecessary use of x rays. . . .

Government insurance programs such as Medicare and Medicaid were conceived as a way to provide adequate medical care to people of all ages and income levels. This goal is laudable. Many people are advocating the expansion of these programs into a comprehensive National Health Insurance Plan which would pay for medical care for

25

all citizens. Senator Eagleton of Missouri estimated in 1976 that National Health Insurance would cost taxpayers about $80 billion dollars a year. Many critics of current medical practices feel that such a program would lead to an incredible rise in hospital costs, doctors' fees, and unnecessary diagnostic tests and therapeutic procedures. They feel that nothing short of a major change in the health care delivery system will alleviate the problem.

The problem of how to provide adequate medical care to all people without fostering overdependence, overutilization, and rising costs is a difficult one. One solution which has been suggested is an expansion of the number of health maintenance organizations (HMOs) like the prepaid group-practice plans such as the Health Insurance Plan of Greater New York, and Kaiser-Permanente and the Cooperative of Puget Sound, which are found on the West Coast. These organizations are designed to provide comprehensive medical care including preventive, diagnostic, outpatient, and hospital services to a voluntarily enrolled group of people who pay a fixed fee on a regular basis. The physicians are usually salaried or have an income sharing plan, and the hospitals are usually owned and managed by the plan.

A study of comprehensive HMO plans revealed that they entail a higher degree of consumer control and greater provision for preventive treatment than other insurance programs.[6] HMOs also appear to deliver services at a cost which is lower than that for the same services performed by private physicians on the usual fee-for-service basis.[7]

Because physicians working in prepaid plans are salaried, there is no financial advantage to ordering extra tests or treatments. If anything, HMO doctors might have a tendency to use too few x rays and tests. However, patient participation in prepaid plans is voluntary and those who do not feel they are receiving adequate medical attention will probably seek alternative forms of medical care.

Most of the HMOs operate in industrialized areas where their growth has been stimulated by labor unions negotiating for better fringe benefits. The expansion of HMOs, which currently enroll about 2½ million people, to other areas has been discouraged by the American Medical Association which has a bias towards fee-for-service practice.

Nevertheless, prepaid plans still appear to suffer some of the shortcomings of other methods of insurance coverage. The results of a study completed in 1975 indicate that all other factors being equal, those who have prepaid use medical facilities more extensively than those on a pay-as-you-go system. In this study a group of physicians practicing together were serving prepaid and fee-for-service patients at the same time. The prepaid patients made 100 percent more office visits and 75 percent more hospital visits than the other patients.[8] No judgment was made in the study about whether prepaid patients were overutilizing services or whether perhaps the fee-for-service patients were underutilizing services.

However, these prepaid group practice plans still subject patients to fewer diagnostic tests, hospitalizations, and surgical procedures than other insurance plans. There is no evidence, yet, that the decreased use of medical services by people in the plans has an adverse effect on health. Thus, in spite of the AMA's opposition to them, these plans are worthy of further study and consideration.[9]

Whether or not the health care delivery system is overhauled, the dilemma of how to provide adequate and affordable medical care to all people without suffering from "the tragedy of the commons" remains.

One approach is to limit insurance so that it only covers major or "catastrophic" medical expenses. This limitation cuts down on doctor and clinic visits for minor complaints by making patients pay for routine services. This relieves the poor of unbearable financial burdens. However, this type of plan does not provide any incentive to physicians or patients to cut costs.

The answer to the insurance problem probably lies in a program which strikes a compromise between major risks and comprehensive coverage. However, the success of such a program requires profound changes in the economic and organizational structure of the health care delivery system. These changes in the system must serve to protect patients from unnecessary medical visits, diagnostic evaluations, and surgery.

CHAPTER 4
RULING OUT THE
WORST AGAIN AND
AGAIN

Always think of the worst. This usually provides a dilemma for the physician . . . because to rule out the worst with certainty often means an extensive and expensive investigation which will likely turn out negative. But the physician is always haunted by the responsibility not to miss something important.

> G. Timothy Johnson, M.D.
> *What You Should Know About Health Care Before You Call a Doctor!*

Leaving No Stone Unturned

Most people consider lab tests and x-ray procedures a vital part of a proper medical diagnosis, but without definite rules specifying the number of tests needed for a correct diagnosis, a patient can't tell when a physician or dentist is ordering too many or too few tests. It is likely in our current climate that too many tests are being conducted. Although the fear of malpractice suits and the convenience of ordering tests covered by insurance stimulate overuse of tests, the major cause of excessive testing is the belief that the possibility of missing a diagnosis should be avoided at all costs. Physicians often feel a great responsibility not to miss an important diagnosis, and it is natural that they should. This process of extensive testing, known as a "workup," is designed to "rule out" the possibility of various improbable but potentially serious illnesses. One technician who wrote me about the conduct of workups remarked:

> A medical workup on an inpatient will often include x rays of the chest and gallbladder and an IVP [kidney and ureter], as well as a barium enema and GI series. A lot of these procedures are fishing expeditions. The patient usually has non-specific complaints of one sort or another so a workup is ordered. I feel that a careful diagnosis of an ailment before an x-ray examination would cut the number of unnecessary exposures drastically.

There is tacit understanding that when an individual's health is involved, no expense or inconvenience is too much to bear. Consequently, patients are rarely consulted about the time, expense, or risks involved in checking out their complaints. This lack of consideration can literally wreak havoc on people's lives. One middle-aged woman was hospitalized in 1974 because of an allergic reaction. She complained that she was "forced to have a full set of body x rays—skull to toes. I was so ill that I had no way of protesting this—it was simply forced." It would be difficult to determine from this letter whether the x rays were appropriate or not. The woman's physician may only be guilty of failing to communicate with her about the need for the x-ray studies, but there are times when examinations seem uncalled for because they bear no apparent relationship to the patient's complaint. This is part of the "it's easier to order x rays than to think" syndrome.

Another woman, admitted to the hospital for a urinary tract infection, reported to me that her doctor had decided to consult with a blood specialist. This specialist proceeded to order x-ray examinations of the upper and lower gastrointestinal system, the breasts (mammograms), liver, and spleen. At this point the patient became angry:

> I told the specialist that if I needed x-ray examinations they should be of the urinary tract. Finally both my doctors agreed. After these x rays they found nothing wrong with my bladder but reported that I had a gallstone and requested gallbladder x rays. I refused these since I had never had any symptoms of gallbladder problems. I then asked to be discharged, and neither doctor wanted to discharge me until I threatened to walk out. . . . Later another doctor informed me that I had no gallbladder problems.

Sometimes x rays are taken before consultation with a physician, just for convenience. I received a letter from a patient who had no family physician. She entered a hospital emergency room with chest pains and throat discomfort. The patient was immediately x-rayed. Then she saw an examining doctor in his office two blocks from the hospital who gave her a diagnosis of indigestion caused by nervous tension. The doctor told her "the x rays weren't really necessary, but since I had been at the hospital he had x rays taken so he wouldn't have to send me back there."

Sometimes the process of leaving no stone unturned can lead to cruel treatment of people who can no longer be helped. A grandmother reports:

> . . . having decreed that my little three-year-old grandson was doomed to die within weeks of cancer of the throat, the doctors in charge nevertheless ordered extensive x rays of his head. . . . Even the technicians were horrified at the extent of the x-raying—57 at one time. He had already had an extensive series of x rays.

She asks if there is anything that can be done to "stop these merciless practices in . . . hospitals—after which they send enormously exorbitant bills."

Although consumers are not always in a position to judge which x-ray examinations constitute a proper part of a diagnostic workup, they can question those examinations which are clearly taken for convenience, those which are futile, and those which are apparently unrelated to their complaints or symptoms.

Physicians who appear to be taking a random or shotgun approach to diagnosis should be questioned. We as patients ought to request that the purpose of each proposed diagnostic test be explained to us. We need to be informed about the likelihood that the examination will yield valuable diagnostic information or rule out a condition. We need to know how much time and expense is involved in each proposed examination as well as what risks it entails. Finally, we need to be asked on the basis of our understanding of the situation whether we want to submit to these examinations or wait and see how our symptoms develop.

31

A New Set Each Time

Physicians and dentists have a penchant for ordering their own set of x-ray studies even though a patient may have just received the same studies at the request of another practitioner. People with problems which remain undiagnosed or difficult to treat find themselves consulting different specialists, all of whom order independent x-ray studies.

A woman wrote that she was involved in a boating accident in the summer of 1974. She injured her back, and was admitted to the emergency room of the local hospital where x rays were taken. Her doctor reported that the x rays showed no spinal injury and that she should feel better in a couple of weeks. She reported:

> I trusted his diagnosis and let four months go by in excruciating pain before I saw another doctor, this time a chiropractor. He made a series of x rays which revealed a broken coccyx and some jammed vertebrae. After several months of unsuccessful therapy and more x rays, I was referred to another chiropractor who insisted he must take new x rays of his own. He administered more therapy, but to no avail.
>
> An osteopath I tried did not x ray, but he gave up on therapy after eight sessions. He referred me to an orthopedic surgeon who made more x rays, and said there was nothing he could do for me that time could not do, and to come back in six weeks when he would take more x rays to see if there was any improvement. I didn't bother to go back since there had been no improvement in over a year. But his x rays revealed that I had indeed broken a vertebra.
>
> I became greatly frustrated since it seemed that each doctor was finding something different.
>
> As a last resort I went to a neurosurgeon, a very reputable one, who admitted me to the hospital for a myelogram [an x-ray examination of the spinal cord after the injection of a contrast medium]. I was also given several regular x rays, including two sets of chest x rays because the first set did not "take". This episode ended in my undergoing a laminectomy and spinal fusion. By the time you read

this letter I will have had another series of x rays as part of my post-operative checkup.

My main discouragement lies in the fact that I had so many unnecessary x rays in such a short period of time, and they were all concentrated in my lower back which involves the reproductive organs . . . my husband and I would like to have another child soon.

Physicians are not the only health care professionals who bounce patients around between specialists repeating x rays every time. One woman wrote on behalf of her husband who lost a filling:

Our family dentist, Dr. F., took an x ray and filled his tooth. He told my husband if he had any more problems he should have it pulled by Dr. G. A week later my husband went to see Dr. G., who took another x ray of the same tooth and sent him to Dr. S. for root canal work. He took an x ray of the same tooth. He in turn sent him to Dr. P. who specializes in treating pyorrhea, who took a complete set of x rays of the entire mouth. All of these x rays in one week!

Why couldn't they use the same x rays over and over by letting the patient carry them from doctor to doctor like they do in Europe? Once you pay for your x ray in Europe, it's yours to keep.

Repeated x-ray examinations are especially frustrating to patients when it becomes apparent that a physician does not even bother to read previous x rays. An individual who had this experience in 1974 sent me the following description:

. . . I had an attack of acute diverticulitis and ended up at an emergency hospital. Before they would give me anything to kill the pain, they insisted on taking four x rays and a blood sample. I submitted because I needed help. However, the four x rays turned into eight because the technician neglected to check the condition of the machine before taking pictures—and I got exposed four times extra.

. . . From what happened to me next, I am sure the doctor in the examining room never even bothered to study the x rays before treating me (but I won't go into that—it just spelled more inefficiency).

33

They finally administered the shot to relieve the pain and sent me home and told me to see my doctor the next day.

I reported to my doctor's office the next morning with a drooping right eyelid and a raised right eyebrow. I don't know if this was caused from the x ray or from the shot they gave me. My doctor never told me and I was too sick to care at the time. My doctor took about 30 x rays. I had asked if he couldn't just use what the hospital had already taken and he merely said he needed more. I doubt that he ever sent for the hospital x rays. I remember asking the doctor why he was taking x rays of the chest when the trouble was so obviously in the bowels—he said he had to check everything before making a diagnosis of my problem. After a week the doctor sent me to another lab to have barium x rays, which again amounted to more exposures (I lost count). At least they finally found out what was wrong with me but I hated being exposed to all those x rays and wondered just how much was really necessary.

Although repeated x rays are sometimes needed to follow the progress of a condition, the risk and expense of very frequent follow-ups in patients sometimes seems unwarranted. A mother who had been living in Texas writes:

My first experience with x rays was when my five-month-old daughter swallowed a small safety pin. . . . I took her to a small nearby hospital where she had an x ray to see where the pin was. It was in her stomach.

Another x ray was taken the next day to see if the pin had moved. It had not. The third day another x ray showed that the pin had not moved, so I was sent to the county hospital with the three x rays.

The doctor at the county hospital looked at the x rays I had brought but wanted his own. They took one from the front and one from the side. A few minutes later I was told the x rays didn't turn out so two more were taken. Again I was told these didn't turn out. Someone finally told the person who was taking the x rays that she was using the wrong adjustment. Two more x rays were taken. The doctor looked at them and said the baby would be admitted. I was sent back to the x-ray room. I protested. Why more x rays? I was told it was the rule that when someone was admitted they had to have x rays, so the baby had two more.

. . . They observed her for a week and gave her more x rays during that time. I have no way of knowing how many or how many mistakes they made. . . .

This series of seemingly endless repeats is especially disturbing because the baby was being subjected to abdominal examination, which entails more radiation than most other medical x-ray procedures. In addition, a five-month-old baby is about ten times more sensitive to radiation damage than an adult would be.

The same mother had another baby girl, a newborn, who was also subjected to extensive x-ray examinations. This time the mother was living in California. The year was 1968. She reports that:

The baby was born blue and wouldn't breathe on her own. When she was finally breathing it was too fast, her heart beat too fast, and every once in a while she turned blue.

So of course . . . the x rays started to see if she had pneumonia or heart trouble.

The next day she was sent to a larger hospital in Los Angeles where more x rays were taken.

For the next few months she didn't seem to gain much weight. They kept insisting she had heart trouble even though all tests were negative. Then they insisted on taking x rays of her long bones to see if she was a dwarf, and then they ordered a skull series consisting of five x rays.

Finally, they wanted to inject a dye and take several x rays of her kidneys. By then I was fed up. . . . I took her out of the hospital against their advice and never went back. Now she is eight and still small for her age but she seems very healthy.

Knowing what I do now and not being as shy, if I could only go back, I would never have allowed all those x rays.

The curtailment of the personal rights of servicemen also allows for the use of repeated x rays, in a military setting. A young Viet Nam veteran serving in the Marine Corps who read my x-ray guide for consumers described the following situation:

> During my tour of Viet Nam I incurred a herniated disk. For approximately one year after my symptoms first began, the military medics thought I was faking just to get out of the Marine Corps.

> During this time, they shipped me back and forth between four different naval hospitals until they did a laminectomy at Portsmouth Naval Hospital in Virginia. Before this time, they would constantly force me to take more and more x rays of the lumbosacral area [lower back]—no shield, of course.

> You may find this hard to believe, but I would estimate that they exposed me to close to 200 x rays! Since you are government property and have few rights, you have no right to refuse them!

Carelessness and mistrust of previous films are not the only reasons for retakes. A sick woman wrote about being referred to a new "multi-million-dollar hospital" where she was subjected to numerous retakes of a throat examination. In between these retakes she overheard a teacher advising an x-ray technician student on how to correct mistakes made on previous films. She wrote, "I would again and again have to hold still in terribly uncomfortable positions . . . . I was furious that sick people would have to go through such an ordeal. It was about 70 minutes, instead of the 20 minutes it should have been . . . and it is apparently dangerous."

Retaking x rays for teaching purposes also occurs at certain dental schools. A woman visited her dentist with a temporary crown which had broken and received a set of full mouth x rays. When her dentist informed her that she needed $750 worth of work, she decided to see if the work could be done more cheaply at a local dental school. She reports:

> I brought along the x rays taken by the first dentist, who was on the school's faculty. The choice was mine, I was told, but no work was done on patients unless students could first x-ray their teeth for practice. That, I was told, is the school's standard and apparently irrevocable policy.

So, a student took a full set—24 frames. . . . However, to get those useable frames, a total of 35 x rays were taken. I was counting and getting upset. . . . For this I paid $18, on top of the $25 two weeks earlier.

Several dentists I have talked with feel that a full set of dental x rays should only require 18 films.

When patients ask why repeats of recent examinations are needed, they are often given vague answers like "I need more." I have discussed this problem with several radiologists who feel that these repeats occur for several reasons. Physicians and dentists tend to mistrust x rays taken at a facility with which they are not familiar. They claim that many films are badly exposed or improperly developed, and that the records which should accompany the films are missing or incomplete. Some physicians seem to feel that having films *they* ordered will be better protection in case of a malpractice suit. If a practitioner is following the progress of a condition, he may want consecutive x-ray films taken to verify his medical treatment.

The radiologist with whom I spoke felt that practitioners were often reluctant to spend the extra time needed to obtain previous films. One obvious solution to this problem would be to allow a patient to carry his films with him when visiting different specialists or facilities. Most x rays are properly labeled with information about the patient, facility and date. However, when such information is missing or incomplete, the x rays themselves are of little value to a new physician.

Unfortunately, many physicians, chiropractors, and dentists regard the x rays they have ordered as solely their own property or to be released only to other professionals. One woman, suffering from pain in her shoulder, went to a chiropractor who took x rays of her shoulders. She comments:

A few weeks later when I attempted to pick up these x rays in case I wished to see another doctor (I had no intention of going back to this one), he refused to release them unless the request came directly from another doctor, even if we were moving to another town. Why must we always go through this hassle to get our own x rays? We have paid handsomely for them, and they are of no use to the doctor who took them when we decide to change doctors.

37

Here is another patient's story of what happened when she attempted to gain possession of her dental x rays before moving to another city:

> I asked a periodontal dentist for some x rays I had just had taken for some follow-up work which had not been completed. Several calls asking to have them mailed were ignored and I finally made an appointment to talk with the dentist himself. He told me flatly that the dentist owned the x rays (my regular dentist had taken these particular ones) and he would not give them to me. Nor did he let me have the ones he had taken in previous work or some taken by another dentist who had referred me to him. (Mysteriously, he mailed them several weeks later with no comment!) His manner was firm, annoyed and patronizing. I just left, feeling very put down.

Who actually owns a set of diagnostic x rays—the prescribing physician or dentist, the hospital, or the patient? In the past there has been no clear-cut answer to disputes over the ownership of x-ray films. Often local courts have ruled that patients who pay for an x-ray diagnosis are paying primarily for their consultation with the physician, dentist, or radiologist who reads the films.[1] When the x-ray films are considered by a court to be a tool used by the professionals with whom the patients consult, they belong to the x-ray facility or prescribing physician. However, a precedent for the patient's ownership of x rays was set in a Potomac, Maryland court by a woman who sued for possession of her dental x rays.[2] She had received several sets of x rays from dentists attempting to diagnose a long-standing tooth problem. She was referred to a root canal specialist who took his own set of x rays. After the x rays were taken the woman decided that she did not want to be treated by the specialist. In order to avoid extra radiation and expense she asked to take the x rays with her to the next dentist. When the specialist told her he would only release the x rays to another qualified professional, the woman sued for possession of the dental x rays, and a Maryland district court judge ruled that if she paid for her x rays she was entitled to them.

Even though it is not always clear who has the legal right to keep x-ray examinations, patients should try to gain possession of appropriate x-ray studies before consulting a new physician or dentist. Professionals may be willing to have copies of films made when they do not want to release the originals. Patients should also submit to x rays only at well-equipped and well-staffed establishments which have a reputation for conducting high-quality x-ray examinations.

It is helpful for patients to keep a complete record of what x-ray studies have been conducted in the past. Such a record should include information about the type of study, its purpose, the location of the x-ray facility, and the date. Patients who keep such records will then be able to inform their new physicians and dentists about the existence of any previous x-ray examinations which may be of interest to them, and insist that they be used whenever possible.

Chronic Backache

About one out of every three adults suffers from an aching back, and most doctors claim that next to respiratory diseases, back pains cause more office visits than any other medical complaint.

Most backaches are chronic or recurring and are located in the lumbar or sacral regions of the lower spine. The problems of diagnosing and treating low back pain are summarized in the recently published *Encyclopedia of Common Diseases:*

> Doctors persist in a never-ending battle over the basic cause of low back pain. Some adhere to the classic assumption that low back pain is almost always the result of an injury or faulty habits of posture; others espouse newer research which suggests that chronic back trouble is due to a degenerative disease that everyone has to some degree; still others, specifically psychiatrists, believe that a great deal of the chronic back trouble is psychosomatic in origin. The treatments themselves are just as varied in nature as the suspected causes, ranging from simple aspirin to surgery.[3]

The exact diagnosis of the cause of lower back pain can be elusive and prohibitively expensive. The follow-up treatment is often time-consuming for both doctors and patients.

Attempts to diagnose and treat chronic lower back problems provide a major source of x-ray exposure. About six million x-ray examinations of the lumbar spine or lumbopelvic region are performed each year in the United States.[4] Unfortunately, these x rays of the lower back are among the medical procedures which provide the highest doses of radiation to critical organs. Large amounts of x rays to the bone marrow in the pelvic region increase the risk of leukemia, and the exposure of the reproductive organs of potential parents may cause damage to future generations. Although it is almost impossible to shield the ovaries of women receiving lower back x rays, it is usually possible to provide shielding for the male testes. Nonetheless, x-ray operators often neglect to do so.

Among the most tragic letters I received about repeated x rays were those from patients searching for relief for intense lower back pain. For example, an industrial worker from New Jersey injured his back on the job. He went to a general practitioner who ordered a complete set of spinal x rays and then referred him to an orthopedic surgeon who took his own complete set of x rays but could not diagnose his problem. In desperation, the patient referred himself to a chiropractor who took full body x rays—still no diagnosis. He then returned to his general practitioner for another examination and was referred to a neurologist. Each time he visited a physician or chiropractor he received several x-ray examinations of the lower back, plus chest x rays which were required by hospital x-ray departments.

The man's workman's compensation and medical insurance were totally depleted, so he and his wife filed bankruptcy and returned to West Virginia to live with his mother. Since he was still disabled and suffering from acute backaches, he began another round of consultation with specialists *and* x-ray examinations. Each practitioner refused to send for or use previous x-ray studies. By the time this man wrote me he had been subjected to more than nine complete sets of lower back examinations and six required hospital chest x rays in the six months after his injury. He was still in pain, and swore that he was never going to visit a medical doctor again.

Another correspondent described an even longer and more tragic search for the cause of her lower back pain. After injuring a leg muscle in 1966 she developed spasms in her back and sought medical attention from a series of physicians:

I was aware that x rays are dangerous, but . . . I consented to have

40

x rays whenever doctors ordered them just to get a quick diagnosis. I also saw a chiropractor who took x rays.

Almost every time I had x rays, technicians made errors and asked me to wait to see if the x rays were OK. Many times a second complete series was taken within minutes.

At a major clinic where I had a back x ray, a radiologist thought he saw something on my ovary. I was asked to come in for another x ray. One year later, I was asked to come in for another exam, which showed that the cyst had developed and I needed surgery. Six months later when I was being prepared for surgery, I had more x rays.

Before surgery the pain in my back and legs continued, and I wanted help with this problem. I saw two orthopedists within a few months' time, each of whom ordered x rays.

My back and leg pain continued and so in 1973, six months after my surgery, I consulted a doctor at another medical center who ordered x rays of my chest and back. In 1975 I saw another orthopedist who ordered more x rays.

From 1971 to 1975 I had at least nine series of x rays of my back and stomach, plus repeats by technicians. . . . Never did a doctor or technician ask when I had a previous x ray.

In March 1976 I saw a neurologist who prescribed drugs and told me to call back in 10 days. During my 10-year search for relief, no doctor ever prescribed physical therapy as a possible way of helping me. The most frequently prescribed treatment was "take some tranquilizers, let me know how you are in 10 days." Today I am in rather serious condition, having constant spasms in my back and leg muscles.

Perhaps there are few solutions for such people. Some patients, like the woman who just described her case, have concluded that physicians order too many useless diagnostic tests, resort too frequently to drugs, and don't emphasize physical therapy and exercise enough. On the other hand, some physicians feel that patients expect too much

41

from their physicians, instead of taking a more active role in maintaining their own health. For example, Dr. Hugo Keim, a leading orthopedic surgeon, stated on a CBS news program in 1975 that:

> They [patients] have a job where they don't exercise at all. And so, after all these years of abuse, they now come to a doctor with back pain, but expect a pill or a shot or a laying on of the hands or something that's going to suddenly transform them into the youth that they were twenty years ago, and this doesn't always work. And they don't realize that what really is the key to the whole situation is that they must change their way of life.[5]

Each back problem is unique and it is impossible to cast the blame for suffering entirely on patients or entirely on physicians. However, if the letters I received are at all typical, it is clear that some physicians must shoulder the blame for unnecessarily repeating high dose lower back x rays.

Chiropractors and X rays

Chiropractic treatment was developed around the notion that the principal cause of disease is misalignment of the vertebrae in the spinal column. These spinal misalignments are known as subluxations. Although many chiropractors now recognize other sources of illness, such as bacteria and viruses, they still emphasize the spine as the major source of illness. Thus, it is common for chiropractors to conduct spinal x rays even when a patient's complaint seems unrelated. A woman wrote that:

> I had been having bursitus in my left shoulder and had gone to a chiropractor. He took four x rays which covered my complete spine —front and side, top and bottom.

The major form of treatment used by chiropractors is spinal manipulation, sometimes known as "the laying on of the hands." Chiropractors receive less training than medical doctors and are held in low esteem by them. Nevertheless, patients who are disillusioned with conventional medicine, especially those with back problems, often visit chiropractors.

In my Consumer's Guide to Medical and Dental X rays, I did not mention anything about the x-ray practices of chiropractors. One dentist was incensed that I hadn't dealt with the "problem":

> My main criticism of your Consumer's Guide is your glaring omission of the number one source of totally gratuitous x-radiation of the American public—chiropractic x rays.

I received packets of anti-chiropractic material from two organizations of physicians—one in Canada and one in the United States. The deputy director of the Health Protection Branch of the Canadian Department of Health and Welfare called me about chiropractic x rays. He was disturbed because a group of chiropractors in his country were using my guide as evidence that only physicians and dentists abuse x rays—not chiropractors.

There is very little solid statistical data available on the ways in which American chiropractors use x rays, and for this reason chiropractic x rays were not discussed in the original guide. Chiropractors are popular in both Canada and the United States, and a 1971 survey of the *Journal of Clinical Chiropractic* indicates that more than 10 million x-ray examinations were being conducted by United States and Canadian chiropractors annually. At least two million of those were the type which irradiates the body from the skull to the thigh, including the thyroid, bone marrow, and reproductive organs.[6] These organs are considered among the most susceptible to radiation damage. It appears from the letters I've received that chiropractors are no more innocent as a profession than physicians and dentists when it comes to the inappropriate use of x rays.

Several physicians who wrote to me charged the United States government with covering up or withholding information about chiropractic x rays. But spokesmen at the U.S. Bureau of Radiological Health say they have less data about chiropractic x rays simply because there are fewer chiropractors. The two major studies on U.S. x-ray exposures conducted by the bureau involved obtaining information about the x-ray examinations received by a random sample of the American population. Because of the relatively small number of chiropractic x-ray examinations in the U.S. (about 1 percent of the total), there were too few of them in the random sample to draw firm conclusions about the practices of chiropractors as a whole.

Those people who view chiropractors as quacks obviously feel that all their x rays are unnecessary. Radiologists have pointed out that the quality of the image on x-ray films of chiropractors who take x rays of the whole spine is usually poor. This poor quality is attributed to the fact that the body is not uniform in thickness over the whole spine so that some parts of the films have to be overexposed and some underexposed.

A representative of the American College of Radiology, an organization which represents about 11,000 radiologists, commented that "probably fewer than half of all chiropractors have x-ray machines . . . but those who do tend to perform their own examinations with the aid of persons usually lacking the x-ray technologist's qualifications."

A disgusted consumer wrote that "one chiropractor in our area opened a new office and advertised as an added attraction—free x rays." The repeated use of spinal x rays by chiropractors has been stimulated by the Medicare/Medicaid requirement introduced in 1973 that chiropractors must submit x rays on an annual basis for their patients before they can be eligible for government reimbursement.

Since mutual respect between chiropractors and physicians is lacking, chiropractors are probably even less likely than medical doctors to send for x rays ordered previously by a physician.

Chiropractors are subject to all the same rules of good practice for the operation of x-ray equipment as physicians and dentists. Thus, chiropractors, like medical doctors, should shield the reproductive organs whenever possible, request and read previous x rays, and use proper techniques to assure a high quality x-ray image.

CHAPTER 5
PATTERNS OF
OVERUSE—ROUTINE
EXAMINATIONS

Many physicians, and I am among them, feel that the traditional 'annual physical' has been overpromoted. . . . Most physicians would agree that, in the absence of any specific complaint, chronic problem, or family history that needs checking, the annual exam is unrewarding.

G. Timothy Johnson, M.D.
What You Should Know About Health Care Before You Call a Doctor!

I'm having the last lap of my annual check-up. My doctor is, to say the least, thorough! This time is a barium check of the intestines. Just had all the works (which, by the way, I watched on closed circuit **TV**!). I *saw* my own intestinal tract and watched the barium travel around in it.

Anonymous patient, 1976

Routine examinations are those in which people without any complaints or symptoms of illness are advised to submit to diagnostic tests. These routine examinations involve regular medical and dental check-ups as well as mass screening programs in which large segments of the population are encouraged to undergo testing to help health officials

47

determine the source of potential health problems. Several types of common routine examinations involve the use of x-ray studies. These studies include the x-ray examinations associated with regular physical and dental examinations, free mobile unit chest x-ray screening for TB and other respiratory diseases, as well as the x rays required for employment or hospital admission.

Screening programs and regular checkups are usually conducted on the grounds that people with unsuspected illnesses can have them diagnosed and treated at an early stage, and, indeed, some cases of unsuspected disease are uncovered. However, regular examinations cost money, divert resources from other health programs, and in the case of x rays, involve some degree of risk. The key question about various screening programs is not whether any disease is diagnosed, but rather whether enough cases of disease are uncovered *and* successfully treated to justify the costs and risks associated with the examinations.

Programs involving routine diagnostic tests are often introduced before the benefits, costs, and risks are properly weighed against each other. One reason for this is probably that the benefits seem immediate and tangible to those conducting the programs while the risks are not so readily apparent. Another reason for the promotion of screening programs is the health care establishment's bias towards detection and treatment rather than true prevention. In his book, *The End of Medicine,* Rick Carlson observes that "In the lexicon of cancer researchers, prevention means early examination, Pap smears, and diagnostic x rays. It does *not* mean eliminating carcinogens in the air and in food products." [1]

Some public health officials currently believe that under most circumstances, no x-ray examination should be performed unless a physician or dentist has interviewed or examined a patient to determine a special need for that type of x-ray examination. Unfortunately, there is no clear agreement on this among doctors, [2] and a number of routine examination and screening programs do expose patients to x rays before they have had any contact with a physician or dentist.

Two of the most controversial routine screening procedures involve the use of mammograms to detect breast cancer, and the use of x-ray pelvimetry to determine the pelvis sizes of pregnant women. Since these types of routine x-ray examinations present special problems to women only, they will be treated in the next chapter.

The Annual Physical

Dr. Richard Spark, associate professor of clinical medicine at the Harvard Medical School, feels that the regular physical examination is a waste of time and money.[3] He observes that several criteria must be met if regular screening tests or examinations of patients without symptoms of illness or complaints are to be effective. The test must have a high degree of diagnostic sensitivity, the doctor must be able to use a course of treatment which will correct the abnormality or arrest the disease that is discovered, and the patient must comply with the treatment. The regular diagnostic tests that Dr. Spark feels are of proven value, using the stated criteria, include those for hypertension, cervical cancer (Pap smear), and breast cancer (mammography) for women over 50. A few other examinations which are useful to conduct on a regular basis for selected groups of patients are mentioned. These include testing for RH incompatability in pregnant women and serologic syphilis tests for sexually active individuals. Dr. Spark feels that "most of the remaining components of the health screen, when evaluated critically, have, for one reason or another, been found inadequate." [4] Doctors Donald Vickery and James Fries present essentially the same point of view in their Consumers' Guide to Medical Care.[5] Medical customs die hard, and Dr. Ralph Green, a Chicago pathologist, was quoted by *Time* magazine as saying the following about annual screening: "There is tremendous money involved: internists, hospitals, and many clinics derive a lot of income from this myth in American medicine." [6]

Routine x rays to detect lung cancer serve as an example of a screening test which is not sensitive enough to allow for effective treatment of patients who do not yet have symptoms of disease. A group at the Medical College of Pennsylvania screened men over the age of 45 with chest x rays every six months for 10 years. Initially, none of the men had any symptoms of lung cancer. Over the course of the program, however, lung cancer was discovered in some of them, and they were given immediate and sustained treatment. But only 8 percent survived five years. This is identical to the survival rate observed when cancer of the lung is detected in people having symptoms of that disease. Thus, even when performed every six months, the chest

49

x ray does not detect lung cancer early enough to allow effective treatment.[7]

As new screening techniques are developed, they must be subjected to constant evaluation using the criteria we have discussed. In addition, the possible risks and expense of proposed screening tests should be carefully weighed against the recognized benefits.

Yet, many physicians are in the habit of automatically ordering x rays for patients during routine checkups without justifying the need for them. One woman observed that "many doctors who make a practice of this operate their own office x-ray machines, and there is an obvious economic reason to add x rays to their standard list of procedures."

Routine Dental Examinations

The dentist who requested that I have a routine checkup every six months used to have his hygienist take several bitewing films before he even looked in my mouth. In a bitewing examination, the molars are x-rayed using small paper-covered films with a wing-like paper flap that the patient bites. This kind of routine examination for cavities is practiced by many dentists in the U.S. today.

It is difficult, if not impossible, to figure out what guidelines dentists are following in developing their procedures. And, there seem to be no uniform standards by which dentists determine how often x rays are really necessary. The American Dental Association recommends that:

> The dentist's professional judgment should determine the frequency and extent of each radiographic examination. Determination of the number of film exposures involved in a radiographic examination should be justifiable in terms of the expected yield of diagnostic information.[8]

A committee appointed by the Environmental Protection Agency developed a somewhat less vague recommendation for federal health care facilities by advising that ". . . dental radiographs be taken only after a dentist has examined the patient and established by clinical indication the need for the x-ray examination." [9] One dentist wrote that the dental school where he was trained recommended that 20 x-ray films be taken annually with four bitewing exposures at the end of six months. Another dentist, who was representing a state dental association, wrote to inform me that the current guidelines of the American Dental Association are as follows:

> Full mouth radiographic surveys should be made only every three to five years, bitewing surveys should be limited to one a year, unless a patient has a high decay rate or other problems which require more frequent observation.

I was not able to find this guideline in published form, but this dentist went on to reassure me about the situation, ". . . please be advised that the dental profession is concerned with radiation exposure to patients and is definitely doing something about it." However, I have seen no evidence that the dental profession is working toward uniform guidelines for routine x-ray examination.

In the absence of clear guidelines, it appears that many dentists in the New York area are opting for x-ray examinations as an integral part of routine checkups. A telephone survey of 500 dentists yielded the following statistics about the practices of dentists in the New York City area: 89 percent of the dentists ordinarily include a full set of x rays (whole mouth) in the patient's first visit. Forty-one percent of the dentists repeat x rays of at least part of the mouth during routine reexaminations every six months and another 46 percent repeat them every year.[10]

The actual practices of dentists vary a great deal, but it is also clear from a number of letters I received from all parts of the United States that the practice of taking periodic x rays is distressingly common. I noted with some surprise that each dentist has a different time period between examinations, and differing numbers of films are exposed. One mother reported that "my children and I have yearly bitewing x rays of our teeth at our annual exams." Another parent writes:

> Last fall I started using a new dentist. He said he'd be working in the dark unless he took 24 x rays which I agreed to.
>
> For the six-month checkup, he now wants to take two more x rays. Since I had no cavities before, nor do I have complaints now, I protested.

51

He laughed as though this was a charming idiosyncrasy of mine, and said soothingly that in that case he'd use a lead apron on me.

Yet another set of procedures and professional attitudes were detailed by a woman who became a patient of a new young dentist. She described her second checkup visit:

> Without saying a word to me, his assistant immediately began to take x rays of my entire mouth. Having had this done on my first visit also, I sat there thinking that this (and the fluoride treatment) was to be the usual procedure for the rest of my life, or the life of my teeth, whichever came first.
>
> When the dentist finally decided to put in an appearance, I calmly told him that I felt these practices weren't needed. Much to my surprise he became furious with me and said he didn't need that kind of "hassle" and that I had two choices: I could do it his way— or I could leave his office.

Another patient reported that:

> I had a dentist who took x rays like they were going out of style. I never thought about it at the time, thinking it was essential. This dentist also seemed to charge more than other dentists in our area. We switched dentists and during the past five years our new dentist has never found it necessary to x-ray any of us.

The following tale from a Florida man suggests that for a person with a healthy mouth, dental x rays during regular checkups may be totally unnecessary.

> I have been living for over 40 years in South America and I have had dental checkups twice a year. My teeth were well taken care of although luckily none of my dentists had x-ray equipment and did not see the need to send me to a hospital, which was the only place where an x-ray picture could have been made. This experience shows how unnecessary routine x rays are.
>
> When I came to Miami with no signs of trouble, a recommended dentist insisted on x rays. When I returned to his office he told me that I needed no treatment and had no new cavities. Six months later I had my next checkup in West Palm Beach. The dentist asked to see the pictures taken in Miami, had a look at my teeth and then

asked why the cavities shown in the pictures had not been taken care of. Whom was I going to believe? Well, I had the 'cavities' (two) filled. This dentist, who fortunately did not make x-ray pictures every six months, left after a few years.

When I visited the next dentist (without any signs of trouble) and he said that he would have to take a few x-ray pictures, I asked in a polite way whether it would be possible to make the checkup without x rays. The dentist reached for my hat, put it on my head and said: "Maybe you can find yourself a dentist who will do that."

The dentist I am seeing now at least is making only one x-ray picture per visit. To reduce the danger, in order to have a better chance, I am having now only one checkup per year.

If dentists are using proper techniques and up-to-date equipment, the small risks associated with routine dental x rays are probably less significant than the extra expense to which patients are subjected. Nevertheless, in my investigation of the dental x-ray situation, I have been unable to find any comprehensive scientific studies on the costs and benefits of routine screening. Therefore, patients are confronted with a multitude of different dental x-ray practices and offered no clear, scientifically supported reasons for submitting to routine dental examinations. This means that, at present, patients are free to decide for themselves what constitutes good dental procedure and then choose their dentists accordingly.

In the absence of clear evidence that routine dental x rays are beneficial, it seems reasonable for patients to save time and avoid the expense and potential risks associated with such x rays. I personally would avoid placing myself in the care of any dentist who conducts periodic x-ray examinations without first checking the condition of my mouth. Furthermore, if my dentist felt an x ray was indicated, I would request that he or she give me an adequate explanation of why the x ray is needed, based on my previous history of the results of the visual examination.

Hospital Chest X rays

It is common practice for hospitals to require chest x rays of all out-patients visiting their x-ray departments, regardless of what other x-ray studies have been ordered. In addition, hospital in-patients are usually required to have chest x rays before they can be admitted.

Hospital x-ray departments perform more chest examinations than any other type. Although 37 million chest x-ray examinations were being performed annually in hospitals by 1970,[11] there is no firm information about how many of these were conducted on a routine basis. However, the results of a series of hospital interviews conducted by the New York Public Interest Research Group (NYPIRG) indicate that the number is enormous.[12] Of the eight hospital administrators queried in the New York City Borough of Queens, four have pre-admission procedures which normally include a chest x ray for patients who enter the hospital. Another three hospitals do not include chest x rays as part of their normal pre-admission procedures, but interviewers were informed that doctors normally prescribe chest x rays for their entering patients anyway.

Patients who are scheduled for surgery usually require a chest x ray in order to be sure they can safely receive general anesthesia. However, this is no excuse for routinely x-raying every patient who enters the hospital. Most hospitals that automatically x-ray incoming patients claim to be screening primarily for tuberculosis, which is a highly communicable disease that could spread to other patients. However, this does not seem logical in terms of what is now known about tuberculosis and its detection. The incidence of tuberculosis is currently at a low ebb in most areas in the United States, although the disease is still relatively widespread among people living in overcrowded, poverty-level surroundings. Dr. John DeHoff, Health Commissioner for the city of Baltimore, commented in a 1976 newspaper interview that "if you have someone with no signs of lung trouble on physical examination and with no complaints about their chest, it's been shown in the medical literature that chest x rays do no good." Thus, even though the city of Baltimore has the highest per capita incidence of tuberculosis in the United States, the Health Department there does not advise the use of routine chest x rays. Instead, they recommend a simple tuberculin skin test. According to Dr. Arthur Levin, a physician and author of *Talk Back to Your Doctor:*

> Tuberculosis screening is more safely and more accurately done with the TB skin test than by chest x ray. The TB skin test is a good

example of a screening test which should be more widely used. This test can be done by a doctor or nurse in seconds. The patient can "read" the results at home and call the doctor if the test is abnormal (redness or swelling at the test site) . . . the TB test costs virtually pennies to perform.[13]

In my opinion, the physicians throughout the United States who are responsible for developing hospital admission policies should reconsider the mandatory hospital chest x-ray policies in light of current evidence about their value and in light of the availability of other less harmful or expensive testing procedures.

Many patients who wrote to me objected to these routine hospital chest x rays and were angered by the vague response they received from doctors and hospital employees when questioning them.

A man who entered a hospital near his home in 1975 for an operation on his hand was given a pre-admission chest x-ray examination. Barely two months later he entered the same hospital for the removal of a skin cancer and was forced to have more chest x rays. When he complained vigorously, the technician told him, "Either you get one or you are not admitted. It's hospital policy!"

A woman entered a hospital, also in 1975, with a urinary tract infection. After she was sent for her chest x ray, she wrote about her exchange with her technician:

> I said, "Why? My problem is the urinary tract, not the lungs." I was told that it was hospital policy, that I might have TB and not know it, and that patients coming in must be x-rayed, in order not to endanger other patients. I said, "You have visitors . . . in patients' rooms every day and do you ask them for chest x rays?" The technologist was very disturbed with that question.

Another woman took her six-week-old baby son to the hospital for a hernia operation. She reports that:

> Upon admission he was given a front chest and side chest x ray with a lead apron over his lower body. After the x rays were developed, one was blurred because the baby moved. It was retaken.

> Because he had sniffles, he was not operated on. Two weeks later he was again admitted to the hospital and over my strong protests, was

given two more x rays. This time I had to remind the technician to put the lead apron on him. So there he was at the age of two months with five routine x rays.

Why are these x rays taken in the first place? The only answer I have ever gotten is "hospital policy."

This tale is typical of many I received in which hospitals refused to accept recent chest x rays and insisted on taking new ones instead. A young medical student who wrote to me offered the following defense for the "hospital policy":

People admitted to a hospital are there because they are SICK (or else they shouldn't be there). You know well that cardiovascular and respiratory diseases represent an extremely large portion of morbidity (illness) in the U.S.A. Routine chest x rays on admission to a hospital are part of a complete physical examination, and highly valuable.

There seems to be little or no support for this contention that the routine hospital chest x ray has much value. For instance, when the New York Public Interest Research Group polled hospitals in the New York area, it found that none of the hospitals in the survey had kept records on the number of pre-admission procedures or on the number of positive findings. It is obvious that these hospitals have not evaluated their policies on the basis of the value of these x rays to their patients.[14]

It is well known that x-ray departments are a major source of income for hospitals, and hospital administrators probably have little incentive to take a long hard look at the wisdom of their routine x-ray policies. This is a shocking situation. Even if the added expense and risk of unneeded chest x rays were ignored, these hospital policies add a substantial volume of work to the already overburdened hospital x-ray departments. This extra, unnecessary load adds to an atmosphere in which technicians working under tremendous time pressure make mistakes or treat patients in a rude manner. How many technicians have forgotten to shield the reproductive organs of their patients? How many times do technicians take blurred or improperly exposed examinations?

A valid scientific appraisal of the real benefits and costs, in time and money, of routine hospital pre-admission chest x rays is badly needed. Although such an appraisal cannot be conducted readily at a local hospital, individuals, as well as local and national organizations, can contact some of the organizations which are concerned about x-ray practices. (A list of these organizations is included in Appendix E.) If enough voices of concern are raised, one or more organizations or agencies may undertake a study.

Meanwhile, there is also a need to encourage constructive changes in hospital policies at the local level. Instead of the blindly applied pre-admission chest x-ray requirements, I recommend that:

1. Hospitals not require routine chest x rays of any patients unless they have complaints or symptoms which would require such x rays, or unless their work or home environments predispose them to respiratory diseases.

2. Whenever a patient's complaints, symptoms, or environment warrants an x-ray examination, hospitals obtain and read any current x rays which may yield the required information *before a decision is made to order new ones.*

Individuals as well as members of local civic clubs or organizations can find out more about hospital practices in their own communities. They can work with physicians and hospital administrators to affect needed changes in local policies. I strongly urge readers of this book to do as much as possible to reform the policies of any local hospitals which make a practice of conducting routine chest x rays or repeating examinations unnecessarily.

Employment X rays

X-ray examinations are often conducted for the legal protection of employers. The most common of these examinations consists of the routine pre-employment lower back (lumbar spine) x-ray series for workers doing heavy lifting. Some employers assume that an x-ray examination of the lower back can predict the chances of an employee developing a job-related back disability. But some investigators have found evidence that this is not true.[15]

Although a recent conference on pre-employment physical examinations [16] concluded that routine x rays of the lower back might be valuable in assessing the degree of already existing disabilities, clear evidence supporting this contention was not yet available. Unfortunately, lower back or lumbar spine x rays involve higher exposures than most other commonly used diagnostic examinations.

Another common pre-employment examination is the chest x ray required by certain states for job categories like nursing, teaching, and food handling. One problem here is that mobile van x-ray units are often used to take pre-employment screening tests, rather than conventional clinics or hospitals. The x-ray machines commonly found in mobile vans are of the photofluorographic type. In this type of device, the x-ray beam excites a fluorescent screen instead of exposing photographic film. The fluorescent screen is then photographed, usually with a 35 millimeter camera. These photofluorographic x rays are not usually as high in quality as conventional (radiographic) x-ray films, and they often expose patients to more radiation.

Unions and other organizations representing employees subject to routine x rays can work towards the reform of regulations imposed on them by employers as well as by local and state governments requiring such x rays, especially if they are conducted in mobile vans with photofluorographic equipment.

Many of the people who wrote to me about being subjected to employment x rays are concerned about their rights. One man who entered a hospital for hernia surgery wrote that:

> At the hospital I was given a front and then a side chest x ray. . . . One exposure was mis-positioned and had to be retaken. At the time I felt this examination was unnecessary. However, when I returned to work my employer required me to sign a permission for a physical examination and I was again subjected to chest x rays.
>
> I suppose I could have objected and been told I wasn't rehired. What would my rights to refuse have been?

I also received an inquiry about rights to refuse employment x rays from a young nurse:

> I am a 24-year-old practical nurse and have had a required chest x ray every year for the last five years. Please tell me if I am justified in refusing a chest x ray upon the resignation of my

hospital job. It is the law, apparently, that everyone must have an x ray when they leave.

The only reason I was given for having to have it done is so that if something should show up in the near future they won't be held responsible. In that case, I am not endangering anyone else's health so I see no reason not to refuse.

It is not likely that I have any upper respiratory disease. Previous x rays have not shown anything, and I try to eat nutritiously, and get plenty of rest, fresh air, and exercise.

Unfortunately, in the absence of state laws prohibiting the use of diagnostic x rays without the prescription of a physician or dentist, employees appear to have very little legal recourse. If the employer decides that x rays are a job requirement, then an individual who refuses may lose his job.

Residents of states like New York may be in a good position to refuse employment examinations since there is a law in New York that x rays should not be taken without a doctor's prescription. However, in spite of the law, NYPIRG found that chest x rays are still a common employment requirement in the New York City area. They are so common, in fact, that at least 38 companies in New York City own and operate medical x-ray units for personnel. Of the companies surveyed in this group, 86 percent indicated that they require chest examinations for certain employees.[17]

Another example of the misuse of x-ray examinations is the executive physical examination. Routinely prescribed by the management of some companies for top executives, the executive physical may subject these employees, among other things, to x-ray examination of the spine, kidney, and upper and lower gastrointestinal tracts (GI series). The GI series and lower spinal examinations are among the highest risk examinations of all. This is because the lower trunk is thick and requires a more intense beam of x rays to obtain a good image on the film. In addition, the reproductive organs and a lot of red bone marrow, which is susceptible to leukemia, are in the primary x-ray beam in the lower trunk area.

Although passing an executive physical seems to give many executives a sense of well-being, there is evidence that the inclusion of high radiation dose routine x-ray examinations is not justified in view of the risks.

Mobile Unit X rays

For many years private health agencies have offered free mass screening chest x rays to the public. Countless individuals, without the recommendation of a physician, have subjected themselves to x rays in mobile units. Since chest x rays taken in mobile units often involve significantly more exposure than conventional x-ray machines,[18] risk becomes a more important factor. Although mobile units may have been at one time a necessary part of a viable public health program, the virtual eradication of tuberculosis from most areas in the United States has made screening programs essentially useless. Thus, in 1972 the Bureau of Radiological Health, the American College of Radiology, and the American College of Chest Physicians recommended in a joint statement that mass chest x-ray surveys of the general population for TB and other heart and lung diseases be discontinued.

In spite of this statement, a few local tuberculosis associations still encourage screening by urging people to have x-ray examinations of their chests without visiting their physicians first. The states where mobile unit programs were still in existence in 1976 include Pennsylvania, California, Ohio, and West Virginia. (Some of these remaining programs are targeted for the diagnosis of black lung disease in coal miners.)

A survey in Pennsylvania conducted in 1974 revealed that 27 local agencies conducted 207,000 mobile unit chest x rays.[19] Many of these agencies did not follow up on the results of their programs. Almost half of the agencies surveyed did not even know if any lung or heart diseases had been detected in their programs. Furthermore, those medical conditions that were detected by the chest x ray and required medical attention only showed up in about one case in a thousand. In spite of their knowledge of the joint statement recommending that mobile unit chest surveys be discontinued, only three of the 27 private Pennsylvania health agencies interviewed were planning to phase out their programs.

Fortunately, mobile unit programs have been phased out in most regions of the United States. Anyone who does run across a mobile unit is well advised to avoid it. If you are worried about your heart or lungs, you should visit a physician first. Then if a chest x ray is prescribed, go to a good hospital facility which has conventional equipment rather than submit to the higher dose photofluorographic equipment often found in mobile vans.

In conclusion, then, I encourage individuals who have symptoms or complaints to consult their physicians or dentists. Diagnostic x rays may be beneficial to such individuals. But remember that there is mounting evidence that many types of diagnostic x rays and other tests do not effectively result in improved medical care for patients with no symptoms. Thus, you should avoid the expense and potential risks associated with routine x-ray examinations unless there is good evidence that they are worthwhile.

CHAPTER 6
SPECIAL PROBLEMS
OF WOMEN

"Why," I once asked a woman friend who called me for advice, "didn't you ask your doctor about that during your visit?" "I don't know," she replied. "I was so nervous, all I wanted to do was get out of there!"

Arthur Levin, M.D.
Talk Back to Your Doctor

Women consulting a physician often have an overwhelming urge to flee from the office, prescription in hand, without asking any questions. Unfortunately, the medical profession is still dominated by men, and the patronizing attitude of many doctors towards women reduces the chances of effective communication being established between them. Even well-educated, competent women can become meek and submissive when dealing with their physicians. Women sometimes feel that doctors are quick to label them as "neurotic," merely prescribing tranquilizers for their complaints.

Unfortunately, the special problems encountered by women undergoing diagnostic x rays involve the areas of the body which are traditionally more difficult to talk about with physicians. Many women have written to me for advice about undergoing mammographic x rays for the early detection of breast cancer, about x rays and pregnancy, shielding for their ovaries, and required hospital x rays after childbirth. In this chapter I will discuss the special hazards which x rays pose during pregnancy and as a result of irradiation of the breasts during screening for breast cancer. This discussion is especially important because women often find it difficult to talk openly to male doctors about these topics.

Breast Cancer and Mammography

The average American woman has one chance in 14 of developing breast cancer some time during her life. Breast cancer is the leading cause of death in women between the ages of 39 and 44, and more women die of it than any other type of cancer.

These stark facts about breast cancer led the National Cancer Institute and the American Cancer Society to introduce a comprehensive nationwide screening program in 1972. Known as the National Breast Cancer Detection Demonstration project, this screening program provides free examinations for women over 35 without symptoms of breast cancer at 27 regional centers (See Appendix F for a listing of the centers). These examinations of the breast include palpation, which involves feeling the breast for lumps; mammography, which is a special x-ray examination usually involving two or three x-ray films of each breast; and thermography, a harmless technique which can detect slight differences in temperature between normal and cancerous tissue. This relatively expensive screening program was initiated in the hope that mammography and thermography would aid in the early detection of breast cancer. About 250,000 symptom-free women between the ages of 35 and 74 have been screened since 1973. About half of these women were between the ages of 35 and 49. Until the summer of 1976, women over the age of 35 were being encouraged by the American Cancer Society to present themselves for breast cancer screening on an annual basis.

Most women consider their breasts an essential part of their femininity, and the fact that even breast cancers which are not fatal often result in disfiguring surgery and radiation therapy make this disease particularly frightening. When the wives of both the president and vice-president of the United States underwent surgery for breast cancer in 1974, thousands of concerned women throughout the country flocked to physicians' offices and screening centers for mammographic x-ray examinations.

The assumption of those who set up and operate the screening centers is that early detection, made possible by mammography, will result in more effective treatment and a reduction of breast cancer deaths. Unfortunately, however, the fact that the female breast is very susceptible to cancer also makes it more sensitive to radiation-induced cancers.

A 1965 study revealed that women exposed to repeated fluoroscopic x-ray examinations of the lungs, used during a type of treatment for tuberculosis that was popular in the 1930s, had an increased chance of developing breast cancer.[1] In fluoroscopy, the x-ray image is projected onto a fluorescent screen so that the motion of the patient's organs can be observed. The old-fashioned fluoroscopes used in these lung examinations involved much more radiation exposure than examinations using modern x-ray equipment and photographic film. Although it is impossible to prove that any specific case of breast cancer resulted from these fluoroscopic x-rays, I was saddened by a letter I received from a woman who had fluoroscopic x-ray examinations for tuberculosis as a child. This woman began to have periodic fluoroscopic examinations in 1930 when she was 12 years old. She received additional periodic examinations in a TB sanatorium for about three years. As an adult, she re-entered a sanatorium and had even more fluoroscopic x-ray examinations between 1940 and 1947. At age 39 she developed cancer in the right breast and had a radical mastectomy. When she suggested to a doctor that the cancer might have been caused by the numerous x rays and fluoroscopic examinations, she received a "disdainful negative response." In 1975 at age 58 she had another radical mastectomy in the left breast to eradicate a cancer.

Since fluoroscopic examinations of the chest involve considerably higher doses of radiation than conventional x-ray films, it is quite possible that the incredible number of examinations she received resulted in her breast cancers. Fortunately, most of us will never be exposed to that much medical radiation.

However, the sensitivity of the female breast to radiation-induced cancer raises the issue of whether the use of mammography as an annual screening tool for women who have no symptoms or family history of breast cancer is worth the risk. A woman who is contemplating a mammographic x-ray examination may have a small breast tumor which could be detected and treated before it would be discovered in a physical examination. But she is also running the risk that she may develop a breast cancer induced by the radiation between 15 and 30 years after the x-ray examination.

Since the chance of developing breast cancer increases with age, a group at the Bureau of Radiological Health performed a risk-benefit analysis to determine at what age women without any symptoms of breast abnormalities or family history of breast cancer should begin annual mammographic screening.[2] They concluded that women should not begin annual screening until age 50. First of all, it appears that most women do not develop breast cancer until after menopause. And second, women in their fifties are not as likely to live long enough to die of radiation-induced breast cancer that could appear in 15 to 30 years. These women, then, would benefit by receiving an annual mammogram. According to the Bureau study, the risks for younger women of developing cancer from the radiation absorbed during an annual mammogram outweigh the benefits of detecting cancers that may already be present. On the whole, women under 50 would not benefit by annual x-ray screening for breast cancer.

Dr. John C. Bailar III, editor of the *Journal of the National Cancer Institute,* also raised serious questions about the wisdom of continuing a program of routine screening for all women over the age of 35. Bailar was quoted as saying that "a small group of highly placed, intensely dedicated people used all the political levers at their command to start the screening centers before the risks had been fully studied."[3] Because of the concern about the possible radiation hazards of mammographic screening, the National Cancer Institute and the American Cancer Society jointly appointed three independent study groups to investigate the situation. By summer of 1976, two of the study groups had recommended that the National Cancer Institute and the American Cancer Society discontinue the routine use of screening for breast cancer in symptom-free women below the age of 50.

On the basis of the study group recommendations, the National Cancer Institute issued a set of interim guidelines calling for the removal of routine mammographic examinations for women between the ages of 35 and 50 in the government-sponsored breast cancer detection programs. However, the guidelines did not recommend withholding mammography from a woman in the 35 to 50-year age group "if she and the physician agree that it is in her best immediate interest." [4] While the guidelines are expected to influence procedures they are not binding on doctors in general. Moreover, the apparent reluctance of some of the project directors and members of the American Cancer Society to follow the guidelines raises some question about whether the centers will adhere to them.[5] Many representatives of the American Cancer Society are convinced that the yield of unsuspected breast cancers is quite high. A spokesman for the American Cancer Society pointed out that among the 129,000 women under 50 screened since 1973, 223 breast cancers were detected. One hundred of these were detected by mammography alone. Seventy-nine percent of the cancers found in these women were detected before the disease had spread to the lymph nodes, giving a high probability of cure.

The mammography controversy will not be easily settled in the near future because there is not enough solid information about the risks and benefits to draw firm conclusions about the desirability of routine screening for women between 35 and 50 years of age. For example, it is generally accepted that x rays can induce breast cancer at high doses. But we don't know whether or not x rays have the same effect at lower doses.

Even if the magnitude of the risks associated with a given mammographic x-ray dose was well understood, observers are not really sure how much dose women are actually receiving in typical mammographic examinations. The skin doses actually received by patients can vary tremendously from facility to facility. One survey of facilities in eastern Pennsylvania indicated that some machines and procedures exposed women to as much as 50 times more radiation than others.[6] Some of these variations in dose were attributed to the use of different techniques preferred by radiologists at each facility, but some of the higher dose examinations also resulted from the use of outmoded equipment and poor or sloppy practices on the part of the technicians conducting the examinations. However, proponents of regular mammographic screening for all women over 35 claim that the newer equipment and techniques used at the breast cancer detection clinics expose women to a minimum of radiation.

Experts also disagree about the value of early detection. Mammographic screening is only effective if early detection results in successful treatment. But Dr. Bailar claims that there are no studies which demonstrate that mammographic screening in and of itself results in reduced breast cancer mortality.[7] Thus, there is honest disagreement about both the risks associated with mammographic x rays and their value in helping physicians discover and effectively treat breast cancer.

Advice for Women on Mammography

The mammography controversy has left millions of women in a quandary about the circumstances under which they should submit to mammographic examinations. One woman wrote:

> My doctor has told me that he would like me to have a baseline mammography taken as a screening for breast cancer and as a basis for comparison in the future if the need seems indicated. Do you think it is worthwhile or desirable to have this done?

Another woman who wrote to me was 36 years old, and she had already had a non-malignant cyst removed from one breast. She wanted to know if I would advise annual mammographic screening in her case.

These two women who wrote to me were obviously concerned about the possibility that they might develop breast cancer. On the other hand, they were also aware of the potential hazards of radiation and were fearful about receiving mammographic examinations. This same fear was expressed by some of the women contacted in the New York Public Interest Research Group's 1975 local telephone survey cited earlier. Members of NYPIRG asked 145 women the following question about their attitudes: "Do you think that women who have breast symptoms, like lumps in their breast, should have x-ray examinations to check for breast cancer?" Although most of the women surveyed said yes, a surprising 14 percent said no.[8] These responses indicate that some women may be carrying their fear of radiation too far.

Although unnecessary x rays should be avoided, women must not become so wary of the risks involved that they overlook the times when x-ray studies can be of real value. Thus, in light of the guidelines developed in 1976 by the National Cancer Institute, I recommend the following course of action for all women at the present time:

1. Women of all ages should learn about and conduct self-examina-

68

tion of their breasts on a monthly basis (see Appendix F for details).

2. A woman of any age who has some indication of a higher breast cancer risk, such as a family history of breast cancer, a personal history of non-cancerous cysts, or a previous breast cancer, should consult a physician about the desirability of undergoing diagnostic breast cancer tests, including mammography, on a regular basis.

3. *Women of any age with breast symptoms or complaints such as pain, lumps, or discharge should consult a physician immediately.* In such cases a mammographic examination is usually warranted.

4. All women over the age of 50 should receive a complete breast examination, including palpation and a mammogram, on an annual basis. Those who have no symptoms or family history of breast cancer are eligible to visit any of the 27 breast cancer screening centers for a complete breast examination. There is no charge for this screening. (See Appendix F for a list of these centers.)

5. *Women between the ages of 35 and 50 with no indications or family history of breast cancer and with no symptoms or complaints should undergo regular screening for breast cancer which does not include a mammographic x-ray study.* These women are also eligible to go to any of the 27 breast cancer screening centers and receive a breast examination which does not, *if they so request, include a mammographic x-ray examination.* (See Appendix F for a listing of centers.)

6. A woman who is going to receive a mammographic examination should go to a well-equipped facility which uses low dose mammographic equipment. (See Chapter 14, item 5 under "Checking out the equipment and facilities" for more details.)

Just after formulating the do's and don't's of submitting to mammography, I became the first to follow my advice. I am currently 36 years old. My mother and aunt have both had breast cancer, so I became quite concerned when I experienced a mild pain in the area of my left breast. It lingered for two or three weeks, and I promptly arranged to see my family doctor about a mammogram. The doctor diagnosed my pain as a bruised rib—probably the result of an elbow jab from one of my energetic children. The physician and I both agreed that in spite of his "bruised rib" diagnosis my family history of breast cancer indicated that an x-ray study of my breasts would be appropriate at this point.

I was both apprehensive and interested as I visited the Carlisle Hospital for my mammography. I tried to remember all the advice I wanted to give to the readers of this book about submitting to breast x rays. I determined that the machine was a modern xeroradiographic unit which, when properly used, entails a minimum of radiation exposure. I peppered the technician with questions which she attempted to answer. She seemed to be working with care, although she wasted one shot trying to get the exposure factors adjusted properly.

The radiologist who read my films found an unsuspected and still small non-cancerous growth in one of my breasts. My family doctor and the radiologist suggest that I submit to a follow-up x-ray study in three months. Then, after we all do more reading and see the results of the follow-up study, a course of action will be determined.

Needless to say, in spite of the controversy about breast x rays, I am personally grateful that mammography and other diagnostic x rays are an integral part of modern health care.

The mammography controversy is a very real one which will not be resolved until more information is available about both the risks and benefits of x-ray screening for breast cancer. Thus, I would advise all women to keep informed of any changes in recommendations on mammographic screening and act accordingly. The news media and women's magazines usually carry current information about these topics.

X rays and Pregnancy

The practice of exposing pregnant women to diagnostic x rays which place the fetus in the direct x-ray beam is known to be potentially hazardous to the unborn child.[9] A developing fetus or embryo is much more sensitive to radiation damage than an adult receiving the same radiation dose. A study conducted by personnel at the University of California Medical Center concluded that there are harmful effects of x rays to an unborn child at any stage of pregnancy.[10] Early in pregnancy the risk of accidental miscarriage, congenital malformations, and brain damage predominate. After the fourth month the most serious risk is that the unborn child may develop leukemia sometime during early childhood. If it is true that the potential for damage exists in proportion to radiation dose, then those x-ray studies which place an embryo in the main x-ray beam should be avoided if at all possible. Thus, an x-ray study of a mother's abdominal region or lower back should not be conducted unless it is vital to the immediate treatment of a serious condition in the mother or unborn child.[11] Typical x-ray exposures to the embryo and fetus for different types of examinations are listed in Appendix B and vary widely. For example, a dental examination which does not place the mother's uterus in the main beam exposes the unborn child to essentially no radiation, while certain abdominal examinations deliver radiation exposures which are larger than those associated with any other diagnostic procedures.

It has been reported that in 1970, 23 percent of the three and one-half million pregnant women in the United States were exposed to medical x rays.[12] Nine percent of these examinations placed the unborn child in the x-ray beam. The potential damage to unborn children caused by such x rays is disturbing to say the very least, especially in cases where the x rays appear to be unnecessary.

For example, I received a letter from a woman who wrote:

> I had been very sick with a strep throat, and still wasn't feeling well, so I was sent to a clinic for tests. While there, not knowing I was pregnant, I underwent an x-ray examination of the chest and three x rays of my lower spine.

> Two months later I had a miscarriage. My gynecologist said he could not find any reason for my having lost the baby and that I should not have any trouble with future pregnancies.

> The doctors at the clinic said I did not receive a high enough dosage of x rays to cause any internal harm. The more I think about it, the more I believe it was the x rays that caused the miscarriage.

Although there is no way to determine whether this woman's miscarriage was caused by her low back x rays or another problem, no one at the clinic bothered to ask her if she might be pregnant. This is inexcusable!

Part of this disregard for safe x-ray procedures may be due to lack of awareness on the part of the technicians who operate x-ray machines. A former radiologic technician who wrote to me in 1976 said:

> During the time I was working as a technologist, I never heard anyone explain to a pregnant woman the dangers to the fetus. I was not aware of any great dangers, and none of the technicians I worked with were aware of any special dangers. The radiologists never said much about the subject, and I did not become aware of the dangers until after I quit work four years ago.

The New York Public Interest Research Group telephone survey (which was discussed earlier) indicated that women of child-bearing age are not usually asked by either physicians or technicians if they might be pregnant.[13] Of those women in the survey under the age of 50 who had received abdominal x rays, two-thirds reported that they had not been asked about pregnancy. Although this survey was small, it agrees with my over-all impression that physicians often fail to ask the crucial question, "Do you think you might be pregnant?" when they are considering abdominal x-ray procedures for young women. A radiologist once admitted to me that he was embarrassed to ask single women requiring abdominal x rays if they might be pregnant. *It is very important for any woman, married or single, who is pregnant or thinks she might be pregnant, to discuss it with her doctor before submitting to abdominal or lower back x rays.*

And this extra caution during pregnancy should extend to other possibly hazardous agents, like medications. A woman who is or thinks she might be pregnant should tell her doctor about this before submitting to x rays or taking any prescription or non-prescription drug. The key question she should ask is, "Is this x-ray procedure or medicine really necessary right now, or can I get along without it until after my baby is born?"

A woman wrote to me about a situation in which she informed a technician that she might be pregnant before receiving an x-ray examination in her abdominal region:

In 1972 I was given diagnostic x rays to detect ulcerative colitis. This was at the office of a radiologist to whom I'd been referred by a college doctor. I was 28.

Before the x ray was taken, I told the technician that I had recently stopped taking birth control pills and that I couldn't be sure I wasn't pregnant. I asked if that was a problem. She said, ". . . well, we don't *like* to take pictures if the patient is pregnant . . ."— and simply proceeded to prepare me. I asked if I could speak with the doctor afterwards, but was told I would have to speak with the doctor who referred me.

As I discovered later, I was three weeks pregnant at the time. My husband and I were quietly alarmed and tried to get as much information as possible about the dangers involved. We had to communicate with the radiologist's office four or five times in an attempt to get a report on the amount of radiation I received, so that my 'case' could be studied. We were consistently treated like some annoying pests. Finally, we received a technical chart, unintelligible to us as laymen (we'd been promised something we could understand). We made two more calls and were told, on the last one, that they didn't have any more time to spend on us. At that point my husband, who just happens to be a lawyer, threatened to see the radiologist in court if we didn't get the report. We got it immediately.

The other doctors with whom I consulted decided that I did not receive enough radiation to warrant alarm, and that my pregnancy should proceed.

My child, now almost four, was and is beautiful and apparently healthy but I am not naive enough to believe that the x rays haven't done any harm to him. His lifetime will prove that—won't it?

Another mother described the following experience:

About 15 years ago when I was pregnant, I had a lot of back pain. During my seventh month the pain became more severe, so the doctor took several x rays of my lower back. At the time I was young and very bewildered.

> Well, anyway, the very next day I delivered a 4-pound 11-ounce boy.
> . . . . Some years later when my son started school, his teachers
> told me he was mildly retarded. I don't know if the x rays might
> have had anything to do with it or not.

Since about six percent of children who have not ever been exposed to
x rays are born with some defects, it is impossible to know whether or
not this child's retardation was caused by the pre-birth x rays. It certainly seems, however, that the mother's physician might have been
able to wait until after her delivery before prescribing the lower back
x rays.

In 1970 an overwhelming 88 percent of the fetal exposure to
x rays occurred during the last three months of the mother's pregnancy.[14]
Most of these abdominal examinations conducted just prior to childbirth were taken to determine the size of a woman's birth canal or
the location of the baby. Such examinations, known as pelvimetry, are
still being conducted by some physicians on a large number of their
patients, despite increasing evidence of the extra danger x rays pose to
unborn children.

A mother commented that:

> My first child was delivered by a doctor who does an x ray on all
> women in the eighth month of their first pregnancy. He never took
> any internal measurements as he relied on the x ray. Several other
> questionable practices quickly prompted me to change doctors, but
> not until my first child was thus exposed to radiation.

The following letter reveals an irresponsible attitude towards routine
pelvimetry on the part of one obstetrician:

> During my pregnancy, I read voluminously on nutrition, exercise,
> prepared childbirth, etc. One particular book which impressed me
> was one by Ashley Montague which contained a chapter on the
> hazards of x rays to the fetus. Since it was well over ten years old,
> and I had seen another article in a current magazine dealing with the
> subject, I didn't really expect any conflict with my obstetrician, al-
> though I had realized by that time that he was in the best tradition
> of the paternalistic uninformed "woman's doctor." Thus, I was
> surprised and frightened when, after a routine check-up in my

eighth month (fall, 1974), I entered his private office to see him filling out an x-ray form. My mind was immediately filled with visions of breech births, umbilical cords tied in knots, etc. But no, when I asked him if anything was wrong, he assured me that everything was fine . . . that this was standard procedure for all his patients. In response to my "Why?" he repeated the comforting answer which he had given me on several prior occasions, "We've been doing it for twenty years, and it hasn't hurt a baby yet." When I committed the unpardonable sin of questioning this logic, the poor man lost all patience, crumpled the x-ray form and informed me that if I got into any trouble during labor because of this, I was to consider myself on my own.

Comforting thought for a woman due to have her first baby in three weeks. At this point, I decided that discretion was the better part of valor, and found another doctor who seemed to be planning to be there if I needed him.

The attitude of this woman's doctor is typical of that of other physicians who claim that x-ray studies during pregnancy are harmless because they've been prescribing them for years and "it hasn't hurt a baby yet." These physicians fail to recognize that the long-term effects of such radiation are often not apparent until a number of years after exposure.

The need for routine x rays right before childbirth has been questioned by the technicians who perform them as well as the women who undergo them. A technologist who is concerned about the potential danger of pelvimetry wrote:

> I often wondered how many of these exams were absolutely necessary. The exam requires a heavy exposure compared to other abdominal exams because of the fluid surrounding the baby, and there were many times when some of the views had to be repeated. I wonder how a doctor would react if a patient refused this examination because an informed technician alerted her to the possible dangers to her fetus. He'd probably have the technician fired.

A study by a group at the Bureau of Radiological Health headed by Dr. Kevin Kelly, indicates that results of x-ray pelvimetry on an ex-

75

pectant mother very seldom change the course of her treatment.[15] In cases where physicians are contemplating a cesarean section on the basis that the child's head may be larger than the birth canal, pelvimetries are often performed. Dr. Kelly's group found that in most cases, the decision about whether or not to perform a cesarean section did not depend on the results of the pelvimetry. Although the group indicated that further study is needed, their results suggest that except in very unusual circumstances, pelvimetry may well be a useless examination. This is discouraging when one considers that pelvimetry now constitutes the greatest single source of x-ray exposure of the fetus.

I personally would avoid all pelvimetric x-ray examinations unless my physician could convince me that my special condition warranted the risk to my unborn child.

The subject of x rays and pregnancy is not pleasant, but to end on a happier note: I have received many letters from pregnant women who were concerned about dental x-ray examinations or x rays of the chest or extremities. If precautions are taken during these examinations to collimate the x-ray beam properly (limit its size to the area of interest), there is no need to worry about x-ray hazards since the fetus is not exposed to the x-ray beam during these kinds of x rays. Shielding the abdominal area can provide additional protection in some cases.

Shielding the Reproductive Organs

Men, women, and children who are potential parents run some risk of passing defective genes to their offspring if their reproductive organs or gonads are exposed to x rays. If an examination places the reproductive organs in the x-ray beam, lead shielding should be placed in front of them, provided that the shielding does not interfere with the examination. This is easily accomplished in men and boys by covering their testes with lead shields which absorb x rays.

Unfortunately, the female ovaries are more difficult to protect. There are two reasons for this. First, the position of the ovaries varies from individual to individual so that the shield used must be large enough to account for this variability. Second, because the ovaries are near the spine, urinary tract, and bowels, it is harder to shield them without overlapping organs or tissues under study.

Sometimes the ovaries can be unnecessarily exposed to x rays when they are not supposed to be in the x-ray beam. This happens, for example, in chest examinations where the size of the beam is larger than the size of the x-ray film. It is important in all x-ray procedures to look for proper beam restriction, or collimation (see Chapter 14 for more details). In fact, it has been shown that, for those examinations in which the reproductive organs are not in the direct beam, proper collimation or beam restriction is more important than shielding.[16]

Girls and young women of childbearing age should be extra careful to determine a real need for any abdominal or lower back x-ray examination, and they should be sure that the x-ray beam is properly collimated during all other x-ray examinations.

Chest X rays for Maternity Patients

The requirement in many hospitals that all patients have a chest x ray in order to be admitted to or stay in the hospital extends to maternity patients. Usually maternity patients are in active labor upon hospital admission. Thus they often escape the routine "preadmission" chest x rays until after their babies are born. One mother commented that "all maternity patients, including those who are breastfeeding, are x-rayed. It seems like an unnecessary exposure of the mother's milk to radiation."

This postpartum chest x ray often disturbs nursing mothers since there is a natural tendency to feel that the x rays might contaminate their milk. Fortunately, x-ray energy can be absorbed by milk without leaving any radioactivity or changes in the milk that might harm a young baby (see Chapter 15 for more details).

Although milk contamination is not a problem, taking a chest x ray after the birth seems similar to closing the barn door after the horse is out. A maternity patient has already exposed the other patients in her ward to TB before it can be uncovered by x rays.

A colleague of mine who has done consulting with professional organizations on x-ray practices wrote:

> The question of the maternity admission is a particularly acute one because of the sensitivity to the potential spread of tuberculosis in a maternity ward and the desirability of isolating a mother-to-be who comes in with any possible symptom of tuberculosis or any other

respiratory infection. Unfortunately, this is more a public health measure to protect others in the maternity ward than it is for the benefit of the patient who is undergoing the examination. Further, the irregular basis on which women come to the maternity ward makes it often impossible to procure the examination before the woman is processed for delivery and thus, you sometimes find the absurdity of a woman having already delivered being asked to stop by the x-ray department on the way out of the hospital for the examination which protocol requires. Even then, it could be argued that if the tubercular condition were disclosed, it would be possible to track back to the other patients who were in the ward at the same time and to check on them and their infants for possible indications of the spread of the disease. However, this whole notion is more and more strained as the incidence of tuberculosis continues to decline, and we have generally tried to advise against instituting this without a fairly clear-cut clinical or social indication of need.

I urge concerned women to act as individuals and as members of local women's organizations to encourage hospitals to reexamine this apparently absurd policy of requiring chest x rays of all maternity patients.

CHAPTER 7
MALPRACTICE,
DEFENSIVE MEDICINE,
AND X RAYS

From the data we have been able to obtain from interviewing physicians who use x rays and from others who interpret them, we suggest that approximately 30 percent of (the) total x rays ordered are related to the physician's concern for the potential malpractice threats and are not primarily designed to assist the patient.

D. H. Twine and E. J. Potchen
A Dynamic Systems Analysis of Defensive Medicine
(M.S. Thesis, MIT, June 1973)

The Failure to X ray—Some Consequences

By the mid-1970s the tremendous increases in malpractice premiums paid by physicians, dentists, and hospitals became a major cause of skyrocketing health care costs. The cost of insuring providers of health care is inevitably passed on to the consumer, either directly through increased charges or indirectly through health insurance premiums. The increase in premium costs for malpractice insurance is directly linked to a dramatic rise in the number of malpractice suits and to an increase in the size of awards made to successful claimants.

For the past few years the annual increase in the number of malpractice claims has been about 10 percent. This increase is accompanied by a 12 percent increase in the magnitude of the awards made to

81

patients.[1] Because of these increases, it appeared by the fall of 1974 that malpractice insurance was going to become unavailable to physicians in approximately seven states.[2] Although immediate steps were taken by state governments and insurance underwriters to remedy this situation, premium costs took a large jump in states such as California, New Jersey, and Pennsylvania where the incidence of malpractice suits is high. This crisis in premium costs and availability received wide public attention when various groups of physicians went on strike in 1975.

Dr. David Rubsamen, a physician and attorney who specializes in the medical aspects of lawsuits, observes that in the past, essentially all medical malpractice suits centered around alleged harmful or improper treatments (acts of commission). Today, an increasing number of suits involve acts of omission including alleged failure to diagnose or effectively treat illnesses.[3] For legal purposes, a physician or dentist must be able to demonstrate that all attempts to diagnose and treat a condition are in line with locally accepted standards of care. However, some courts no longer restrict standards of care to just the local community. This legal commitment leads practitioners to feel that they should use all acceptable means to discover the cause of an illness.

Failures to order enough diagnostic x rays have been the basis of some malpractice suits. A case was cited by Dr. Rubsamen in which a physician's failure to x-ray was judged as not meeting local standards of care.[4] In December 1968, a college student dislocated his right knee playing touch football. A major artery was severed in the injury but there was no bleeding or apparent symptoms of this because the broken ends of the artery were constricted. The student was examined immediately after the injury by a college physician, an orthopedist, and finally, a general surgeon. Although each physician noted that the arterial pulsations in the right leg were extremely weak, they each suspected that it was due to a spasm rather than a break in the artery. The general surgeon did not feel that an x-ray study of the arteries (arteriogram) was needed immediately. Eighteen hours later, when an arteriogram was finally performed, the severed artery was discovered. Subsequently, the patient's leg had to be removed.

The student filed a malpractice claim. During the trial a vascular surgeon from 400 miles away was brought in to testify. He testified that the severed artery was the main source of blood to the lower leg, and obstruction of its flow for more than six hours creates a substantial risk of gangrene. He also testified that an arteriogram is a fairly simple procedure and since arterial injury was suspected, the x-ray study should have been performed. The witness felt that the failure to perform the arteriogram which led to the amputation of the leg violated the standards of due care. The jury returned a verdict for the college student (plaintiff) of $475,000.

Although the ruling in the case just described was in favor of the patient, suits involving failure to x-ray do not always result in judgments in favor of the patient, even when negligence is involved. For example, a 55-year-old California housewife consulted with her family physician in July 1970 for the treatment of chronic fatigue and a cough.

> She saw him 23 times over the next six months, but her complaints continued and she lost over 60 pounds. Finally, in January 1971, the physician ordered a chest x ray, which revealed a bilateral pulmonary malignancy. The patient died a few weeks later.

> Her husband sued. Three medical witnesses testified that, considering the patient's symptoms, a chest x ray was required by at least the fall of 1970. On the other hand, the attorney for the defendant physician presented two medical witnesses who testified that the cancer was multifocal in origin and therefore inoperable from the beginning. The defense made no attempt to justify the doctor's delay in accomplishing the diagnosis. The jury accepted the testimony of the defense on the origin of the illness and returned a verdict in favor of the doctor.[5]

In this instance, the physician was judged careless for not ordering the x ray, but he could not be convicted because it was also determined that no injury or death resulted from his carelessness. The verdict in favor of the physician was due to the judgment that the cancer was already inoperable. Thus, in order for a physician to be convicted it must be shown that he or she was negligent *and* that this negligence resulted in injury to the patient.

The use of diagnostic x rays is clearly a vital part of meeting the standards of good care in certain medical situations. Thus, it is natural that physicians and dentists would tend to take x rays whenever there is even a remote possibility that they might reveal something.

Defensive X rays—A Dilemma

Ordering diagnostic x rays which, according to current standards of care, may prove beneficial in the treatment of patients is certainly legitimate. However, given the current malpractice climate, it is not surprising that physicians and dentists are ordering more diagnostic tests than ever as a protection against possible lawsuits. The practice of requesting medical procedures primarily to decrease the possibility of a lawsuit or to provide a good legal defense in the event of a lawsuit, rather than to help the patient, is commonly known as defensive medicine.

It is well known that one of the most common forms of defensive medicine involves the practice of ordering unnecessary diagnostic x rays. For example, this practice of ordering defensive x rays is recommended in *A Practical Medico-Legal Guide for the Physician:*

> An excessive number of x rays may be taken, but they are necessary to support the diagnosis and *to protect the physician from a malpractice claim.* A lay jury or judge often considers the examination incomplete without pertinent x rays. The patients themselves frequently share this opinion, often attaching unwarranted importance to radiological examinations. . . . However, some patients object to proposed x rays because of the extensive publicity regarding the dangers of excessive radiation; usually their fears are unwarranted, and with a reassuring explanation they will proceed with the roentgenograms [x rays].[6]

In my original consumer's guide to medical and dental x rays I condemned the use of excessive numbers of x rays for defensive purposes. After reviewing my guide, a radiologist reacted strongly to my statements about defensive x rays:

> Dr. Laws condemns the use of x-ray examinations as a defense from possible malpractice litigation. In view of the number of malpractice cases these days and especially the ever-increasing amounts of successful judgments against physicians, I fail to see how anyone can expect physicians to abandon x-ray examinations of their patients as insurance against legal action. In this particular case the fault does not lie with the physicians, but rather with the legal profession and the courts.

A dentist who also read my guide wrote me a letter presenting his reasons for ordering defensive x rays:

> Over the last 18 years I have taken thousands of dental x rays; however, I do share your concern as well as that of my patients over the excessive use of x rays.
>
> We are unfortunately living in a society where if something goes wrong with treatment (a poor result, a complication, etc.), the doctor will be sued. One of our best defenses against this practice is our x rays.
>
> The two lawsuits against me over the years involved cases where x rays turned out to be very important and in one of the cases, I was criticized for inadequate films. This was a fact I was well aware of prior to surgery, but I wished to save the patient the additional expense and also wished not to expose the patient to any more radiation. Well, as things turned out, I was held liable and it was based on inadequate x rays.
>
> My point is this: until the laws are changed in this country regarding malpractice, all practitioners will continue to over-x-ray just for self-protection and self-preservation. You, no doubt, disagree with this line of reasoning, but until you have had your professional ability publically criticized, suffered sleepless nights, developed diarrhea and indigestion, and have been made to feel like a damn incompetent fool, then don't prejudge us.

In the early 1970s, when I first became concerned with the problems of the overuse of x rays, I knew of no instances of a physician or dentist admitting to a patient that certain x rays were for malpractice protection. Because of the growing public recognition of the malpractice problem, some professionals are now beginning to be more open about taking defensive x rays. Since patients are the ones who receive additional radiation exposure from defensive x rays and ultimately foot the bill for them, they find the admission that certain x-ray examinations are for legal protection very disconcerting.

The following story concerning dental x rays serves as a case in point:

> Having read of the danger of x rays, I asked my dentist to skip the

x rays. I told him I would consider the visual examination to be sufficient.

The dentist said he had to protect himself from lawsuits, and if I didn't want x rays taken I would have to find another dentist. I gave in, paid seven dollars, and the x rays revealed nothing new.

I felt rooked and decided that I'm not going back to that dentist.

One man from California, where the incidence of malpractice suits is exceptionally high, was subjected to 95 x rays of the chest and abdomen in a two and one-half year period—many of them unnecessary repeats. He stated:

> As you know, we have a malpractice problem hanging over our heads in this state. As a result, those medical facilities and private M.D.'s who are still practicing are refusing treatment unless full diagnostic x rays are given prior to treatment—even if you change doctors for the same illness!

Not long ago I received a letter from a woman who was subjected to a battery of defensive x rays after an accident. Her incredible tale involves most of the different types of x-ray abuse discussed in this book.

> In early October I was in a car accident. Considering the aches and pains, x rays were, no doubt, in order. I had x-ray series on an ankle, a wrist, full back, neck, and skull. A tidy bill by the conclusion. In addition I was given the hospital-policy chest x ray.
>
> I was hospitalized in a small-town hospital six days at which point my doctor ordered an ambulance trip to a larger town for more back x rays. This, too, was reasonable since more sophisticated machines were required (not so sure the ambulance was in order though).
>
> I was admitted to the larger hospital, saw the bone specialist, and was sent on down to x-ray. Although all my x rays from the first hospital were sent along (same radiologist at both hospitals), I went through the same x rays as before, except for the chest. The chest x ray was the first thing they wanted to take and I said, "Just a minute now." I told them I'd had a chest x ray just three hours

earlier. I also explained the other x rays taken and already present at the hospital. The technician checked. Her comment? "The doctor wanted these." She proceeded to take the general series again on my back and neck. The others were excluded.

Then I went on to the machine that can take small shots of specific trouble areas: this also seemed reasonable since the general x rays indicated trouble with the fifth lumbar vertebrae.

I was returned to my room. An hour later, I was taken down again. I was told the radiologist wanted more x rays of another section of my back—closer to the shoulder blades. This was a new development. After 20 minutes the x-ray tech said I was done but we'd have to wait until the batch was developed. As I lay there, two techs and the radiologist conversed in the next room.

What I heard made me angry. The x-ray tech had no experience with the machine and had taken shots double the width ordered by the radiologist. A new x-ray tech came in and I went through this entire review again—the pain extreme by that time. I'm sure the pain decreased my patience and increased the tendency to question.

Anyway, I was released from the hospital that same evening. Obviously I received no answers to my questions of "Why are these necessary?"

Since I was not covered by any auto or health insurance the bill for 90 minutes of x ray at the second hospital alone is all mine to pay. This also encourages questions. It becomes easy to think of assumed insurance coverage precipitating excess x rays to pay for big machines. The second hospital is in a town of only 15,000 population. I was one of two patients for the day.

After release from the hospital, I asked my doctor from the accident why so many x rays were required. His response? To prevent malpractice suits and to provide thorough evidence in case of insurance or other individual court action. Ouch and agony! Doctor, not me and not with my body. But how, with one leg numb, could I question or refuse whatever was ordered? As it turns out, the numbness was caused by bleeding around the vertebra fracture. As it turns out, I had in seven days some 40 to 50 x rays. As it turns out, I'll be paying for them for months to come. As it turns out, I still don't have full use of one leg, but can't afford to try another doctor. As it turns out, who do we turn to? Where?

It must be recognized that these situations represent only the patient's view of what happened in each case. However, they do illustrate that patients feel justifiably angry about being exposed to and paying for x rays which have no apparent medical benefits. On the other hand, many physicians and dentists are genuinely worried about the possibility of being sued for malpractice for failure to x-ray patients.

Defensive X rays and the Emergency Room

The hospital emergency room is a place where physicians and technicians often deal with patients whom they have never seen before. When an emergency room is overcrowded, or when seriously injured patients are brought in, the staff in an emergency room must make rapid decisions about how to diagnose and treat a range of illnesses and injuries. The lack of rapport with patients and the crisis atmosphere in typical emergency rooms probably leads personnel to order a large number of unnecessary x rays for defensive and other purposes.

In addition, a study by Robert Brook, M.D. and Robert Stevenson suggests that, in general, the hospital emergency room may be a poor place for diagnosis and treatment.[7] Brook and Stevenson examined the outcomes of treatments for 141 non-emergency patients who were "treated" in the emergency room of a Baltimore city hospital. (In terms of the staff-to-patient ratios, quality of patient care, and evaluation efforts, this Baltimore city hospital emergency room was considered the equal of any in Baltimore or perhaps elsewhere in the United States.) Of the 98 patients who received diagnostic x rays, only 37 were informed about the findings of their x-ray studies. Only 14 of the 38 patients with abnormal x-ray results appear to have received adequate treatment for the conditions revealed by the x rays. If this hospital is a representative one, it appears that in these situations there is little follow-through on the results of the x-ray studies which are ordered by emergency room personnel. This lack of follow-through renders many x rays which are taken useless and thus unnecessary.

88

Because of the lack of rapport with patients and the crises that can occur in typical hospital emergency rooms, such places are also likely to be the sources of a substantial number of defensive x rays.

A man who was twice injured on the job wrote to me about his emergency room experiences:

> In the early fall I bumped and cut my head at work. I probably suffered a mild concussion and stayed home from work for a day. My boss suggested that I have the injury checked before returning to work.
>
> I was referred by a local doctor to the emergency ward at the hospital. The technicians went completely overboard and took an excessive number of x rays. When I questioned the necessity of these x rays, I was treated with disdain.
>
> About a month later I blocked the fall of a fellow employee who was on a ladder and bruised my shoulder badly. I should have known better, but I went to the hospital emergency room again— this time to have my shoulder checked.
>
> Boy, did I get the treatment! They x-rayed my head, neck, chest, arms, etc.
>
> . . . I was getting perturbed and insisted on seeing the attending physician before any more x rays were taken. They finally stopped after 10 x rays. When I questioned the doctor, "Why so many?" he just said, "We have to be sure."

A radiologic technologist wrote to me about her experiences in a hospital emergency room:

> If patients arrived in pain or had bruises or swellings (especially car accident victims or compensation cases), the emergency room doctor would immediately order x rays.
>
> I had requests for entire limbs to be x-rayed. For example, the request might include views of the hand, wrist, forearm, elbow, humerus, and shoulder. An examination of this type can involve as many as 15 exposures, depending on the technician.

In most cases, the injuries were limited to contusions with no fracture. I occasionally questioned my orders when I was ordered to x-ray entire limbs or the whole spine—to no avail. The doctors usually mumbled something about "legal aspects."

Head injuries make emergency room physicians extremely nervous, and it is not surprising that skull x rays are frequently ordered for defensive purposes, in spite of evidence that they are useless in the vast majority of cases. It appears that physicians determine a course of treatment for skull injuries primarily on the basis of the patient's symptoms, rather than on the basis of the skull x rays. Yet, many physicians order skull x rays without even determining a patient's symptoms.

The following emergency room experience of a California woman seems typical:

> Recently I was admitted to the emergency room of our local hospital with dizziness and vomiting. The problem was so obviously centered in my inner ear. The doctor met me with the greeting: "Let's get some x rays. What seems to be the problem?" While I was on the examining table (where he never once looked into my ears), a nurse announced the arrival of another patient. His response was the same: "Let's get some x rays. What seems to be the trouble?"

In a study of 570 children admitted to a hospital emergency room with head injuries, the treatment of only one was altered as a result of skull x rays.[8] Another group studied 1,187 skull fractures in over 4,000 children, and felt that the discovery of skull fracture would not alter the treatment of the children.[9] In a similar project, Bell and Loop [10] studied hundreds of adults and children who were brought to a hospital with apparent skull injuries and subsequently x-rayed. They found that patients observed to suffer from confusion, drowsiness, headaches, visual disturbance, blood clots, lacerations, and swelling, but no other symptoms, rarely had skull fractures. Out of 435 patients with some of the above symptoms who were examined with x rays, only one had a fractured skull. However, among 1,065 patients having additional symptoms, such as unconsciousness for more than five minutes and discharge from the ears, 93 had skull fractures. Again, this indicates that in the case of head injuries, the patient's symptoms rather than x-ray findings are the primary indicators of skull fractures.

Bell and Loop estimated that an average of $7,500 is spent on negative x rays before a skull fracture is detected. They suggested that between $25,000 and $38,000 is spent on skull x rays for each bit of information that in any way alters the treatment of the patient.[11] In short, it is very expensive to find skull fractures using x rays, and in most cases the medical treatment of the patient is not changed by the x-ray results. This is due to the fact that the treatment of the patient is the same regardless of whether or not the skull is fractured. Even though an increasing number of physicians are aware of the fact that the medical benefits of most skull x rays are non-existent, they may feel that such x rays are needed for legal protection in case of malpractice suits. Thus, according to Bell and Loop, the expense of many of these defensive skull x rays constitutes a hidden malpractice premium which would amount to as much as $15 million annually. Moreover, $15 million represents a small fraction of the total amount consumers pay for defensive x rays each year.

Defensive X rays—Costs

A fact-finding commission appointed by the Secretary of Health, Education, and Welfare reported in 1973 that the practice of defensive medicine was one of the most pervasive results of the medical malpractice problem.[12] However, the actual extent to which diagnostic x rays or other procedures are used in defensive medicine is not known, and estimates vary widely. A survey of physicians in California and North Carolina conducted by the staff of the *Duke University Law Review* revealed that about 22 percent of x rays and other diagnostic procedures were conducted primarily for defensive purposes.[13] An American Medical Association opinion survey conducted among members in the spring of 1972 found that an estimated 71 percent of the diagnostic tests they conducted were defensively motivated.[14] The low percentage in the Duke University survey could be a result of physicians being less candid with the law school personnel than they are with personnel from the American Medical Association. These surveys obviously yielded widely different results. Part of this difficulty in determining the "actual" proportion of x-ray examinations ordered for defensive purposes is that practitioners often have mixed reasons for ordering a diagnostic procedure.

However, even if the defensive x-ray procedures constitute only 20 percent of the total number conducted, consumers are paying a tremendous price for the medical profession's malpractice protection in terms of both unnecessary health risks and dollars. It has been estimated that medical and dental x rays cost Americans $4.8 billion in 1975. If 20 percent of the x rays are ordered for defensive reasons, then almost one billion dollars per year are going for x rays which are not medically justified. Since the total awards to patients for malpractice settlements amount to only about $100 million per year, patients are paying ten times more for defensive x rays than they are getting back in settlements from *all types* of malpractice suits. And, in any event, only a small percentage of the $100 million is awarded in cases involving failure to order appropriate diagnostic x rays. Thus, it appears that health professionals may be ordering an unreasonable number of defensive x rays. However, many physicians and dentists have argued that if fewer defensive x rays were taken the incidence of malpractice suits based on failure-to-x ray might be much higher. Thus, the defensive x-ray problem presents us all with a genuine and very expensive dilemma.

The Importance of Doctor/Patient Relationships

Many of the people who write to me about their x-ray problems are frustrated. They have great difficulty communicating effectively with their physicians or dentists. These people often feel hostile and are willing to blame their practitioners for their bad health. A malpractice suit represents not only dissatisfaction with a course of treatment or an attempt at diagnosis; it can also be an expression of anger towards the practitioner or the hospital.

The importance of the relationship between a practitioner and a patient is now receiving more recognition. Physicians who responded to a questionnaire sent out by a professional magazine named "poor communication between physician and patient" as the single most common cause of malpractice suits.[15] The quality of the doctor/patient relationship can be influenced markedly by the personality of either the patient or the doctor. For example, a study of the human relations aspect of malpractice suits dealt with personality traits of suit-prone patients and physicians.[16] A suit-prone patient tends to go into a medical situation with a mistrust of physicians, unrealistic beliefs about all conditions being curable, or the view that physicians should only be paid

92

when they produce results. In addition, the suit-prone patient is often dogmatic and quick to shift the blame to others when things go wrong.

The suit-prone physician is characterized in the study as one who cannot admit to limitations in training or experience. This type of physician often responds to patient dissatisfaction by dismissing the person's complaints as trivial or by ignoring the patient. The suit-prone physician is also described as one who is preoccupied with his or her own image and thus unable to make patients feel less angry, afraid, or depressed.

The special commission appointed by the Secretary of Health, Education, and Welfare to study and report on the factors influencing the rising incidence of malpractice claims, cited the human element as an extremely important aspect of the medical malpractice problem in its 1973 report.[17]

Probably the most important single factor in cutting down on defensive x rays involves improved communication between physicians, dentists, or x-ray technicians and their patients. Even when neither the personnel who prescribe x rays nor the patients have personality traits which render them suit-prone, it is easy for poor communication to lead to hostility between physicians and their patients. Thus, in general, the personnel who prescribe and conduct x rays should take more responsibility for informing each patient about the potential risks and benefits of x rays before a proposed examination is conducted. They should also inform patients that x rays will not always yield results. A physician or dentist should be able to estimate the probability that a proposed x-ray examination will lead to a diagnosis. Patients, too, must do their part by learning more about diagnostic x-ray procedures and other medical tests. They should ask comprehensive questions and not expect an x ray with every medical visit.

The New York Public Interest Research Group noted in a recent study that if steps were taken to improve effective communication for all kinds of diagnostic x-ray procedures, malpractice suits might be significantly less common.[18]

Who's to Blame?

Before proposing solutions to the problems (such as the use of defensive x rays) which result from the rising incidence of medical and dental malpractice suits, it is helpful to review some possible causes of the increase in malpractice suits and skyrocketing malpractice insurance premium costs.

In general, physicians and dentists tend to blame suit-crazy patients, the media, insurance companies, the legal system, attorneys, and uninformed juries for the malpractice crisis.[19] Patients are more apt to accuse physicians and dentists of being negligent, incompetent fee gougers. However, the root causes of increasing malpractice claims and insurance premium costs are so complex that it is difficult, if not impossible, to single out any element of the medical, legal, or insurance system as a primary culprit.

According to the author of *A Practical Medico-Legal Guide for Physicians,* uninformed juries place unwarranted emphasis on the importance of diagnostic x rays.[20] If this allegation is true, then lay juries are indeed contributing to the defensive x-ray problem. Many health care providers feel that lay juries are incapable of separating emotion and fact in evaluating evidence. For example, a pamphlet distributed by the Pennsylvania Medical Society stated that "juries often act on humanitarian feelings and not on proven negligence or misconduct by the physician."[21] However, the Secretary's Commission on Medical Malpractice feels that it is essentially impossible to prove these allegations about the competence of lay juries.[22]

Misinformed juries are undoubtedly a contributing factor in the malpractice climate, at least on some occasions. However, most malpractice claims are settled out of court without juries. Since the media generally give the most publicity to those cases which have great human appeal, the role of emotional juries in malpractice cases is probably exaggerated. I feel there is little justification for citing lay juries as a primary cause of the malpractice crisis.

The complexity of the legal system has also been cited as a major cause of rising malpractice insurance premiums. Malpractice cases usually take two or three times longer to try than other personal injury cases because the required expert medical testimony is so complex. In addition, attorney's fees, court costs, and insurance company overhead and profit consume 84 percent of every dollar spent for malpractice insurance.[23] Only 16 percent of the money is paid to patients in the form of court awards or claim settlements. Clearly, this fact indicates that when it comes to malpractice cases the legal and insurance systems are woefully inadequate and in need of reform. However, the fact that malpractice cases are very time-consuming and expensive to try should, if anything, discourage patients from filing claims. Since the annual rate at which claims are filed is still increasing, these inequities and inefficiencies in the legal and insurance systems contribute more to the rising costs of premiums than they do to the number of excess malpractice claims.

Physicians often accuse the mass media of creating the myth that everyone can be cured if the right doctor is found, and the "medical miracles," especially those portrayed on television, have certainly contributed to greater public expectations about medicine. The notion that patient over-expectation is a source of malpractice suits is supported by Dr. Rubsamen's observations that much of the recent increase in suits involves failures to diagnose or treat illness rather than maltreatment. However, the majority of malpractice suits still are related to patient dissatisfaction with the treatment received.

Juries, the legal system, insurance companies, and the media exist to serve health care practitioners, patients, and the public. Thus none of these institutions can assume primary responsibility for the increasing incidence of malpractice claims. The primary responsibility inevitably lies with health care providers and their patients.

Who is more to blame, physicians and dentists or their patients? It is interesting to note that about 50 percent of all patients who file malpractice claims receive some sort of settlement.[24] Furthermore, insurance carriers feel that about the same percentage of claims has merit.[25] This strongly suggests that health care providers and patients as a group are equally guilty. Changes in the legal system and the image of medicine projected by the media should afford patients more financial equity and reduce the incidence of malpractice suits. But to the extent that improved communication between health care providers and patients can potentially reduce the incidence of actual malpractice *and* reduce the hostilities which lead to unjust malpractice claims, better communication between physicians and their patients is probably the most vital part of the solution to the problems of increased malpractice claims and defensive medicine.

Eliminating Defensive X rays

Physicians and dentists defend the need for defensive x rays in the current malpractice suit climate. There are, however, a number of ways in which the incidence of defensive x rays might be reduced. These include reducing the incidence of actual malpractice, reducing the number of malpractice suits, and convincing health care providers that defensive x rays may not be providing them with significant protection in malpractice proceedings.

Reducing the incidence of malpractice could be most effectively achieved through an adequate system of peer review. In the past, physicians and dentists have been reluctant to acknowledge or exert pressure on fellow professionals whom they felt were incompetent. One professional will seldom give testimony in court which might be publicly damaging to another professional. Dr. Sidney Wolfe, the physician who is director of Public Citizen's Health Research Group, feels that physicians and dentists remain silent because they are afraid that criticizing their colleagues in public will tarnish the image of their profession. Nevertheless, the health professions are beginning to develop and implement systems of peer review. Only time will tell if these are effective.

Reducing the number of malpractice suits is more difficult, but we must strive to reduce the incidence of malpractice claims by improving the communications between patients and their physicians. Several hospitals are taking steps in this direction by adopting statements of patient rights and responsibilities which would improve communication. For example, the statement of rights and responsibilities at Beth Israel Hospital in Boston includes:

- The right to be listened to when you have a question or desire more information and to receive an appropriate and helpful response.

- The right to receive all the information necessary for you to understand your medical problems, the planned course of treatment (*including a full explanation about each day's procedures and tests*)* and the medical outlook for your future.

- The right to receive adequate instruction in self-care, prevention of disability and maintenance of health.

- The right to know who will perform a test or an operation, *and the right to refuse it.*

---

* In my opinion this ought to include information about the costs of proposed procedures and methods of payment.

- The right to leave the hospital even if your doctors advise against it, unless you have certain infectious diseases which may influence the health of others, or if you are incapable of maintaining your own safety, as defined by law.

But communication is a two-way proposition and demands that patients be considerate of their physicians. Some patient responsibilities are also outlined in Beth Israel's statement:

- Being on time for scheduled appointments and promptly cancelling those you cannot keep.

- Bringing information about past illnesses, hospitalizations, medications, and other matters about your health [including x rays].

- Being open and honest about instruction you receive which you don't understand.

- Being prompt about paying bills and providing information necessary for insurance payments.

Patients who follow these sensible guidelines for their rights and responsibilities can gain the respect of health care providers, get better treatment, and help minimize the mutual frustrations and hostilities which might otherwise lead to needless suits.

Additional steps should be taken by physicians and dentists to insure that the public knows more about the limitations of x-ray procedures. The general public, including individuals who serve as lawyers, jury members, or judges in malpractice trials, must be helped to realize that x rays cannot diagnose *all* conditions and that although x rays often appear to provide a form of tangible evidence in malpractice cases, their importance should not be magnified. X-ray studies are not always needed for a sound diagnosis.

Even if the current malpractice crisis is not resolved in the near future, patients should seek ways to encourage practitioners to eliminate defensive x rays. Several factors ought to be brought to the attention of health care providers who are practicing defensive medicine.

1. Most physicians and dentists have never been sued for malpractice, and there was less than one claim per 226,000 office visits in 1970.[26]

2. Malpractice suits which involve failure to x-ray constitute a tiny fraction of the total. In fact, the majority of suits still center around the misapplication of medical treatments rather than a failure to diagnose a condition.[27]

3. A number of courts have ruled that a physician or dentist may be held negligent for failing to take diagnostic x rays, but *only if it is shown that common practice requires the physician to do so in view of the patient's symptoms.*[28]

4. If there is no causal connection between a failure to x-ray and damage to the patient the practitioner will not be held negligent.[29]

In 1976, staff members at the U.S. Bureau of Radiological Health surveyed all of the malpractice claims closed in 1970 which involved either the omission or improper use of x rays. They found that of the 193 claims filed, 50 percent of these claims involved injuries sustained in x-ray departments while x-ray examinations were being performed. Another 24 percent of the cases involved alleged misreading of x rays, while only 24 percent involved alleged failure to x-ray.[30] Unfortunately, once a physician orders an x-ray study, the legal responsibility for its conduct is shifted to the personnel at the x-ray facility. However, it would be ironic if future studies reveal that a physician who orders an x ray primarily to reduce his or her risk of malpractice suit passes an even greater risk of malpractice suit on to the radiologist or x-ray technician who conducts the x-ray examination.

CHAPTER 8
THE WORSHIP OF
TECHNIQUE

In the last few decades, medicine has experienced a flood of technological development. . . . One result is that the physicians have become
purveyors of extraordinarily complex wares.

Rick Carlson
The End of Medicine

The Wonders of X-ray Diagnosis

The faith that technology, properly used, will be able to solve
human problems is deeply held throughout North America and Western
Europe. It is little wonder that we have this faith—jet travel, the
transistor, and open heart surgery are integral parts of modern society.

In medicine, diagnostic x-ray examinations are often the first step
in complex technological solutions to our health problems. The whine
of the x-ray tube anode, the slap of a film cassette, or the glow of a
fluoroscope screen remind us that we are dealing with a sophisticated
modern technology.

New uses for diagnostic x rays are constantly being devised. For
example, a portable x-ray unit was developed in 1966 that was used
by medical researchers to study Egyptian mummies without damaging
them.[1] In December 1975, a *Wall Street Journal* article described a new
computerized x-ray scanner as a "Glamour Machine" emitting super
x rays. The article characterized the scanner as a development "hailed
by doctors as a boon to diagnosis." Other developments, which seem

101

less dramatic but which are just as sophisticated, are announced almost continuously. Improvements are constantly being made in x-ray films and in intensifying screens (phosphorescent materials which are placed next to films to improve their sensitivity). Television-like image amplification devices for fluoroscopic x rays, and the use of xeroradiographic images based on the Xerox process are replacing fluorescent screens and photographic films for certain examinations. Devices which automatically limit the size of the x-ray beam to that of the film are being installed in new medical units. Although they are complex, many of these innovations improve the quality of the x-ray image and reduce patient exposure.

In 1975, the Siemens Corporation of Munich, Germany announced a new device called the "infantoscope" designed especially for x-ray studies on infants. It has specially designed "emplacement contours" to keep the infant in place, and toys dangle from it to hold a child's interest. It is equipped with image amplification to reduce x-ray dosage.[2]

The CAT-Scanner

In conventional x rays, the image obtained on the film is a shadow of everything between the x-ray tube and the film. Many overlapping organs and tissues, sometimes difficult to separate, often show up on the developed x-ray film. Since air, soft tissue, and bone absorb x rays differently, the projections of various body parts on x-ray film have different intensities. However, it is usually quite difficult to see the small differences between normal tissue and diseased tissue even when overlapping organs present no problem. Even with new film, image techniques, and x-ray tubes, certain types of diagnostic information simply do not show up on conventional x-ray films. Now modern technology has changed all that. The "glamour machine" or CAT-Scanner, described in the *Wall Street Journal,* employs a combination of ingenious techniques to detect slight differences between a lesion or tumor and surrounding tissue, regardless of which organs lie above or below it.

The CAT-Scanner (CAT stands for *C*omputerized *A*xial *T*omography) uses a series of narrow, pencil-like x-ray beams to scan the section of the body being studied. The x rays pass through the body and are detected by an electronic device which automatically stays in line with each x-ray beam. Typically about 160 scans are made at one position, then the source and detectors are rotated about one degree and the 160 scans are repeated. When the detector and x-ray tube have rotated completely around the patient, the scan is complete. The thousands of bits of information from the sensors are stored in the memory of a computer. This information is then processed by the computer to reconstruct an image which can be displayed on a television screen or in printed form. A single scan takes anywhere from a couple of minutes to a few seconds depending on the type of scanner and the extensiveness of the examination. In general, the newer models of scanners are faster.

The CAT-Scanner has been lauded by physicians as the greatest advance in its field since Roentgen discovered the x ray. CAT-scanning was originally developed to help visualize brain abnormalities such as hydrocephalus (water on the brain), cysts, tumors, and blood clots. Accident victims suffering from brain hemorrhages often require immediate surgery. CAT scans allow physicians to identify this condition quickly. The scanner also helps physicians to distinguish between strokes caused by bleeding and those caused by blood clots. Although the external symptoms are the same in both cases, the treatments are different. Newer units have been developed to scan all parts of the body.

Although some radiologists claim that the pencil-like x-ray beam exposes patients to less radiation than a conventional x-ray examination, the dose from a complete brain examination is estimated to be equivalent to that associated with a series of skull x rays (two to four rads).[3]

Although the radiation risks for CAT-scanning and conventional x rays may be comparable, the scanning method avoids the risks associated with the injections of contrast dyes needed in some of the conventional procedures. For example, hospitals which have CAT-scanners have drastically reduced the number of conventional brain x rays called pneumoencephalograms. In this particularly unpleasant procedure, spinal fluid is removed from the brain and replaced with air or gas. The gas acts as a contrast medium, enabling radiologists to obtain x-ray images of the brain. The aftereffects of pneumoencephalography include headaches, nausea, and vomiting, and the radiation doses are high (25 to 50 rads).

Another type of brain x ray is angiography, in which a contrast medium is injected into the circulatory system of the brain to make it visible to x rays. The substances used are known to be irritating to the body and the probability of complications and even death is considered uncomfortably high (about 1 percent).[4] Here, too, the patient receives a considerable amount of radiation. CAT-scanning can help to reduce the number of angiograms ordered because it can be used as a preliminary screening tool to determine if the more extensive angiography is needed.

The benefits of CAT-scanners for brain studies are widely accepted, and experts are hoping that the newly developed body scanners will contribute to the early detection of cancers and other diseases in internal organs. There is hope that scanning may detect bone, cartilage, and joint diseases as well as those attacking the blood vessels, heart, and respiratory system. General Electric is designing a scanner for the detection of breast cancer. Dr. Hellier Baker of the Mayo Clinic, who is helping GE, feels that scanning would be of particular value in the detection of breast cancer in younger women whose dense breasts render cancers difficult to find.[5] In spite of the many potential benefits, radiologists admit that the ultimate value of body scanners cannot be assessed without more experience.

As hospitals and medical centers throughout the United States begin to install CAT-scanners, voices of concern about the impact of their widespread use on medical costs have already been raised. Scanners cost from $350,000 to $600,000 each. Furthermore, this purchase price does not include the cost of constructing a lead-shielded room to house the machine, the training of technicians, or the yearly contract fee for maintenance. No comprehensive studies are available on the optimal number of CAT-scanners needed in the United States. Nevertheless, some experts estimate that a ratio of one scanner for every 300,000 to 700,000 people should be adequate.[6] But the interest in scanners has already led to needless and expensive duplication of them in places like Florida and southern California where the ratio is about one per 75,000 individuals.[7]

The average cost to the patient of a CAT brain scan is currently more than $200 and experts estimate that a CAT-scanner with a high rate of use can bring in $2,000 to $3,000 per day. There has been no apparent resistance from health insurance companies to footing the bills for scanning, but if costs rise, the insurers may raise objections.

Another problem is that scanning technology is developing very rapidly. With the first scanners barely three years old, manufacturers are already producing test models of third-generation scanners which will be faster, easier to operate, and cheaper to manufacture. This means that a half-million-dollar investment could very quickly become obsolete. There will be pressure on health facilities which do purchase scanners to earn enough revenue to return their investments as quickly as possible. As more hospitals purchase scanners, economic pressures will induce them to perform scans on more and more patients who don't really need them. This is likely to result in additional increases in medical costs and more unnecessary radiation exposure to patients.

Towards Less Complex Technologies

Most x-ray machines are not nearly as complex or sophisticated as the CAT-scanner. However, the average level of complexity of conventional diagnostic equipment has also grown rapidly in the past few years. This is partly due to advances in x-ray technology which enable manufacturers to produce and sell machines for a wide range of specialized applications.

In 1968, Congress passed the Radiation Control for Health and Safety Act to help protect the public and machine operators from unnecessary exposure to radiation from x-ray machines and other electronic products. The Act authorized the Department of Health, Education, and Welfare to set standards for radiation-emitting devices, including x-ray machines, and to carry out programs to improve x-ray practices. One result was a government standard which required that manufacturers improve the performance and safety of x-ray machines sold after August 1, 1974. The standard applies to major x-ray machine components such as tube assemblies, controls, voltage generators, fluoroscopic imaging assemblies, and the beam-restricting devices known as collimators. In a survey of recent non-compliance actions taken by the Bureau of Radiological Health, federal inspectors found approximately 80 defective components for every 100 dental x-ray machines inspected and 19 defects in every 100 medical x-ray units inspected.[8] These defects included excessive leakage of x rays from x-ray tube housing, omission of appropriate lead shielding around the x-ray tube, spurious radiation after the intended exposure was completed, failure in the collimator control circuitry, improper beam-limiting devices, inadequate x-ray beam filtration, inaccurate timers, and low tube voltages resulting in unnecessary skin exposures. Many of these defects have been corrected and we are indebted to the federal and state inspectors for their thorough work. However, the number of defects found suggests that as machines become more complex, manufacturers are having greater difficulty in meeting minimum radiation safety requirements.

Although the standard can help to reduce needless x-ray exposure, a side-effect has been to increase the complexity of medical x-ray equipment. For example, the standard requires that most x-ray machines have automatic beam-limiting devices, so that an x ray cannot be taken unless the beam size is no larger than the film. This helps to minimize the possibility that a careless operator will use an unnecessarily large beam and thus expose the patient to additional radiation. But the beam-limiting devices are themselves complex, and, like most complex apparatus, they are subject to breakdown. This exemplifies a common situation in today's society in which an advance in technology (in this case the x-ray machine) creates a problem (the hazard of unnecessary radiation exposure), which is solved by still more complex technology (the positive beam-limiting device), which in turn has its own inherent problems (possible malfunction).

One solution to this kind of technological spiral is to improve the competence of the individuals who operate complex x-ray equipment, so that they can assume responsibility for on-the-job quality control and can identify and deal with malfunctions. Another approach, particularly when a shortage of qualified operators exists, is to develop less costly and sophisticated devices which do not perform as many functions but are easier to maintain and operate. The late Richard Chamberlain, a University of Pennsylvania radiologist, did just this when he developed a small portable x-ray unit for use in underdeveloped countries by operators with a minimum of training.[9]

I am not suggesting that sophisticated x-ray machines such as CAT-scanners should not be developed or used, but that their use should be limited to large regional medical centers where they can be monitored and controlled by appropriately trained operators. Patients with special problems could be referred to these centers for further tests when the need arises. Under this policy the majority of hospitals would still have the basic equipment used to perform common examinations.

Widespread use of less complex x-ray equipment is an example of what the British inventor-economist E. F. Schumacher would call an intermediate technology.[10] One guiding principle of such a technology is that the people who use it must be able to understand it completely.

Local organizations in smaller communities are often asked to help raise funds for new equipment and hospital facilities. Readers who belong to these organizations are urged to study the real needs for more sophisticated equipment or facilities carefully before helping to raise funds for them.

Humanizing X-ray Technology

Apparently we have been drawn into a system which renders us more and more dependent on technological solutions to our problems. We are so awed and intrigued by technology, that we fail to see what is happening. Frederick Ferré, a noted philosopher, is convinced that our love of technology is a matter of religion, that technology has become something that is worshipped. Ferré refers to this idolization of technology as "Technolatry," [11] and he suggests that technology should be returned to its proper place in human enterprise—that of an important tool. Ivan Illich speaks of the development of "convivial tools" [12] to replace those large, specialized tools which separate and alienate people from others.

Whether we strive for the secularization of technology, convivial tools, or intermediate technologies, it is important to step back and look realistically at various x-ray and diagnostic systems to be certain that they truly benefit the people for whom they are supposed to be designed. X-ray technology is not to be rejected, but rather, to be used more wisely, more humanely.

107

CHAPTER 9
MANAGING X-RAY
TECHNOLOGY

Every technician is merely doing his job, taking orders, keeping the
machinery running. The crucial intuitions of right and wrong are dulled.
Within the social machine that is the giant bureaucracy of corporate society . . . no part need feel personally responsible, even involved.

Frederick Ferré
Shaping the Future

The techniques which characterize the methods of operation in
modern institutions, including medicine and dentistry, are far more
extensive than those used to operate machines or physical devices.[1] For
example, in a modern x-ray department the actual operation of the x-ray
machine is only a small part of a much larger process which usually
begins with an x-ray prescription by a physician and ends with a radiologist examining the films and reporting his or her findings to the
referring physician. Thus, the effective use of diagnostic x rays depends
heavily on the techniques used to organize and insure communication
between various professionals involved in ordering, conducting, and
reading x-ray films. The x-ray technology system encompasses a complicated set of techniques and procedures which, in ideal cases, function
smoothly and effectively to yield films which provide maximum diagnostic benefit for the patient with a minimum of expense and radiation
exposure.

Technological systems tend to lose reliability as they become more
complex. The modern diagnostic x-ray system is no exception because

as the complexity of x-ray machines and examination procedures increases, organizational procedures necessary to support their use become more and more complicated. Patients often feel that they are wasting their time, money, and good health in x-ray departments because the x-ray machines are difficult for technicians to operate properly, or because the communication among physicians, technologists, radiologists, and patients is ineffective. Hopefully, the discussion in this chapter about how the x-ray system operates and how it breaks down, will help patients understand how to cope with the system more effectively.

Technician Technique—It Matters A Lot

A diagnostic x-ray examination is an involved process which includes preparing and positioning a patient, loading film cassettes, setting exposure factors on the x-ray machine, taking steps to protect the patient from unnecessary radiation, developing the films, judging their quality, and answering patient inquiries. Technicians who are improperly trained or who work under time pressure often fail to complete all the steps required to insure a high quality x-ray film with minimum radiation exposure to the patient.

A mother who had read about my x-ray guide in a health magazine wrote me on the subject of technician errors. Her insight resulted in part from the experiences of her two daughters. One of her daughters was training to be an x-ray technician and the other worked for a company that reclaims silver from discarded x-ray films.

> Do you know how many films are shot that don't come out properly —too light or too dark? Technicians laughingly say, "I fried him that time."

> When a patient is brought in, a "test film" is taken to see if the machine is set properly. Each patient is a different size, shape, and age, and each requires a different setting. After the test film is developed the examination is conducted.

> The technician and students put their test film and other discarded films in "goof boxes." Later they are sold to a company that reclaims the silver. I am explaining this to impress you with the fact that there are enough "goofs" to support a reclamation business in my community.

110

In too many cases, improperly trained or careless technicians cause patients needless inconvenience, x-ray exposure, and expense.

Another mother described x-ray examinations which had to be repeated. Her 15-month-old son received a series of chest x rays to follow the progress of his pneumonia.

> The baby was put into a contraption designed for children so that he would be in an upright position while the x rays were taken.
>
> He screamed as one technician and three trainees roughly strapped him into the device. I was not allowed to stay in the room while the x rays were taken but could hear the conversation. It consisted of the trainees asking if this or that was the correct setting and not sure of the correct exposure. As you may have guessed, one of the x rays was too dark and had to be retaken. The baby had to be put back in the device, again screaming as the x ray was taken.
>
> Since my sister was taking courses in x-ray technology at the time, I spoke to her about what had happened. She told me of similar experiences at her hospital. On one occasion an elderly woman came for x rays. Because of her age there was some difficulty positioning her properly, so the x rays were just taken over and over again. It seems that the trainees were not impressed with the danger of excessive x rays and sometimes use the x-ray machine as if it was a simple camera.

These repeated examinations might have been avoided if the technicians had taken more care in positioning their patients and measuring their sizes to determine initial exposure factors more accurately.

A Florida man who entered a hospital was "forced" to have a routine hospital chest x ray. He reported that "the technician had to take *four* shots before she got an exposure that was acceptable." The patient attributed this to "technician incompetence."

A public health nurse who wrote to me had a similar experience when her eight-year-old daughter had an abscessed baby tooth and visited a dentist. She recounted that "an obviously unqualified dental assistant took five x rays of the tooth, none of which came out, so the dentist took a sixth."

Lack of training or qualifications are not the only reasons for improper techniques. Simple carelessness is also the cause of many poor quality x rays. For example, a woman who had an acute abdominal pain visited a hospital emergency room. She reported that the technician neglected to check the condition of the machine before taking four x-ray exposures. She went on to say, "It was unfortunate that the x-ray technician was being visited by his girl friend at the time—this may have been responsible for his not paying attention to his work. I would blame the hospital's lack of discipline rather than the boy for allowing outsiders to pay social calls during working hours."

Even if the machine settings and the position of the patient are optimal, the film must be developed properly. One letter writer commented:

> I work in a hospital and often see patients have 'routine' x rays taken more than once, many times because there has been some foul-up in the film developing. Patients are never told why their x rays must be repeated and often worry, needlessly, that something is wrong with them. And some worry, too, about excessive exposure to radiation.

Several studies have shown that the primary reason x rays are repeated is that the image on the film is too dark or too light, while the second most common cause of retakes is improper positioning of the patient.[2] Films which are too dark or too light usually result either from failure to choose the proper exposure settings on the x-ray machine or from improper development of the film.

Exams which must be repeated due to technician errors or other technical failures are an obvious source of unnecessary radiation exposure and wasted expense. However, these repeated examinations may only represent the tip of the iceberg, since there is evidence that many of the x rays which are considered to be of "acceptable quality" by technicians (and even by some physicians) are so poor that they contain no useful diagnostic information. Dr. Benjamin Felson, a leading expert in chest radiology, stated in 1975 that about half the chest films used in mining communities to diagnose black lung disease were of unreadable quality.[3] Patients are seldom aware of this source of misdiagnosis as well as unnecessary radiation exposure and expense.

The percentage of films which are retaken during diagnostic x-ray examinations is reportedly between 3 and 6 percent, if the results of recent studies are valid.[4] This would indicate that only about one out of twenty films is being repeated immediately in a typical hospital radiology department. This figure seems amazingly low to me, since almost everyone with whom I have spoken about diagnostic x rays in the past four years and essentially all of those people who wrote me about their experiences have recalled a higher retake rate. For one thing, the studies do not count examinations repeated at a later date because a radiologist or physician felt that the films were unreadable. Then, too, many technicians are probably reluctant to report these mistakes truthfully.

Unfortunately, even when the image on an x-ray film is properly exposed and of good diagnostic quality, the failure of a technician to utilize other standard protective techniques may result in excess patient exposure to x rays. Two major sources of unnecessary radiation are failure to restrict the size of the x-ray beam to the film size or less, and failure to provide lead shielding for the reproductive organs of potential parents.

Many people who wrote to me complained that shielding was not used during abdominal examinations. A number of these people were parents who were concerned about protecting reproductive organs of their children. As I explained in Chapter 6, it is often difficult to shield the ovaries of girls and women. On the other hand, it is almost always possible to shield the testicles of men and boys who are potential fathers. This shielding for males is important because x rays can cause mutations in the sperm cells. Many of these changes, when passed on to future generations, could result in genetically caused illnesses.

Although shielding the testicles is important and desirable in examinations which place them in the primary x-ray beam, it is often forgotten. The Bureau of Radiological Health estimates that the conscientious use of shields for the male reproductive organs on a nationwide basis would result in a 75 percent reduction in testicular dose to the male population.[5] This could lead to a reduction in the number of defective genes passed to future generations.

Although patients are seldom aware of improper beam restriction, it, too, is a major factor in the unnecessary exposure of patients to radiation. Over half of all medical x-ray examinations expose patients to additional radiation because the x-ray beam is larger than the area of the film, with 15 percent of all x rays having a beam area *more than two times larger than the film area*.[6] (Advice on how and when to look for proper beam restriction and shielding is included in Chapter 14.)

In some instances, a combination of exposure settings is used to produce a film which is diagnostically acceptable to the radiologist but causes more radiation to be absorbed by the patient than necessary. In such a case there may be an alternate set of exposure factors which exposes patients to less radiation and results in an even better quality x-ray film. Technicians should take the extra time to figure out the combination of exposure settings that will expose the patient to less radiation while still producing a high quality x-ray film.

The Bureau of Radiological Health, in cooperation with state radiation control agencies, is in the process of conducting a National Evaluation of X-ray Trends (NEXT) survey. In this survey the exposure which an average-sized patient would receive from certain types of diagnostic x-ray examinations is being measured at a large number of x-ray facilities throughout the United States. So far the survey shows a tremendous variation in patient exposure among different facilities performing the same examinations. In response to these NEXT survey results, the Bureau of Radiological Health is developing a quality control program to teach x-ray machine operators how to use better techniques to reduce patient exposure without reducing the diagnostic value of the film. However, with over a hundred thousand machine operators in the United States working in thousands of facilities, it could take years for a quality control program of this nature to have a widespread impact.

The state of Illinois has assumed real leadership among the states in reducing patient exposure to diagnostic x rays by establishing regulations restricting the allowable patient exposure from various types of medical and dental x-ray examinations.[7] The method Illinois regulators use to determine these exposure limits is quite interesting. In 1969, the Illinois Department of Public Health began accumulating data on patient exposures during regular inspections of x-ray facilities. The data was analyzed for each type of examination. If 75 percent of all facilities could obtain an x-ray film of acceptable quality by exposing a patient to less than, for example, 200 mR (millirads), then a rule was developed that no facility could expose a patient to more than 200 mR for that type of examination. (See Appendix A for a discussion of exposure in units of mR.) If the Illinois program is successful, it could prove very worthwhile for concerned individuals and groups to lobby for similar federal legislation.

Improper x-ray exposure, film development, and patient positioning, as well as neglecting to use appropriate collimation and shielding
are all technician errors. The most often cited reason for these errors
is inadequate training for technicians. Within the last few years Puerto
Rico, New York, New Jersey, and California started licensing their
x-ray technicians. Several other states currently have the legal authority
to set requirements for licenses. Congressional bills have also been
introduced to set standards for the accreditation of technician schools
and to require licensing of all United States x-ray technicians.

In 1975 there were approximately 130,000 diagnostic x-ray
machines being used in medical facilities. Fifty-six percent of the technicians operating these machines either had state licenses or were otherwise credentialed by professional organizations. The remaining 44
percent were operating x-ray machines with no licenses or special cre-
dentials.[8] Results of the NEXT survey indicated that approximately 40
percent of the credentialed operators surveyed failed to collimate the
x-ray beam properly while approximately 60 percent of the non-credentialed operators failed to do so. However, there was little difference
in the performance of credentialed and non-credentialed technicians in
terms of the average patient exposure from chest x rays.

Thus, it appears that in actual practice, the performance of x-ray
technicians, whether or not they are credentialed, needs a great deal of
improvement. However, credentialed technicians do somewhat better
than those who are not credentialed.[9] These conclusions indicate that
although a federal licensing program for x-ray technicians may help
reduce unnecessary radiation exposure, it will not accomplish miracles.
It appears that factors such as the total work load in a busy x-ray department and the working relationship between technologists and radiologists in an x-ray facility are also very important. In order to examine
some of these factors it is helpful to understand more about the total
environment in which x-ray technicians function.

The Communication Gap and Passing the Buck

Technicians often feel that they are tiny cogs in a giant wheel,
that they bear no personal responsibility for obtaining high quality x-ray
films with a minimum of patient exposure. They often feel little obligation to communicate with a patient about the conduct of an examina-

tion. One man wrote to me about visiting an x-ray clinic for a diagnosis of pain near his heart. His physician ordered an x-ray series for him. The man reported that the technician took 14 x rays. At this point the patient asked why so many exposures were necessary, and the technician merely replied that "those were the doctor's instructions." Such shifting of responsibility or blame from one individual to another is common.

A technician explained in a letter to me how patients who question the need for an examination were handled. "They were told their doctor ordered it (patients usually accepted their doctor's judgment), or some other excuse to get them to cooperate. Few patients refused examinations because they felt that their doctors and the technicians knew best."

A woman who had some trouble with her throat was also sent to a hospital facility for an upper GI series. She questioned the amount of radiation she would be receiving, and later wrote to me about her experience:

> When the technician found that I had had some experience with x rays he told me that I would get less exposure in the "swallowing function" than he often gave to pregnant women. "Gee, some of them are really thick—and so we have to crank it way up." "What about damage to the unborn child?" "Well—I dunno—we haven't had any complaints." When I pointed out to him that defects appearing when the child was several months old would hardly be traced to prenatal x rays that the parents assumed to be safe, he said, "Look —the doctor orders certain pictures—it's my job to give them to him. He's the responsible physician."

> The doctor orders pictures and assumes the technician will do nothing to endanger the patient in getting them. The technician (like all para-medical people, it seems) endows the doctor with more knowledge and wisdom than any human can possess and assumes the doctor would not order anything that would harm the patient. So the responsibility falls between two stools and the patient, as a result, may get irresponsibly x-rayed.

The patient who wrote this letter is suggesting that the technological system which delivers diagnostic x-ray examinations is operated by individuals who lack communication with each other and that no single individual in the system is willing to take overall responsibility for the efficiency of the process. My own experiences and the many letters I have received from x-ray technicians and patients have led me to believe that the communication problems among prescribing physicians, radiologists, technicians, and patients do indeed constitute a major factor in the unnecessary x-ray exposure which patients receive.

Not long ago, several radiologists were asked in my presence what they would do if they discovered that a woman who came into their x-ray facility with a prescription for an abdominal procedure was pregnant. These radiologists felt that the woman ought to be referred back to the prescribing physician so that this physician could reconsider the real need for the abdominal x ray. However, most of these radiologists said they would not refer the patient back to the physician because he or she might be offended. These radiologists apparently viewed themselves as responsible only for supervising the conduct of examinations. They felt little obligation to question the need for x-ray examinations even if their experience with diagnostic x rays or their greater understanding of radiation hazards would lead them to believe a certain examination was not desirable. Thus a "pecking order" emerges with the prescribing physician at the top and the radiologist next. The radiologist is followed by the technician who is definitely afraid to question apparently useless examinations because the radiologists might be offended. A technician who had performed many apparently useless skull examinations in a hospital emergency room commented that "an experienced staff technician can usually tell if an x ray is necessary or not but few doctors will accept a technician's judgment." This technician requested that her name not be used, and said "I believe many technicians agree with my comments, but I do not feel the doctors would agree."

Another technician who wrote stated that "unfortunately, in some settings, the radiologist is looked upon as the 'almighty' and to question the almighty is unheard of. I have, upon occasion, enjoyed watching a radiologist attempting to teach techniques that he obviously knew nothing about to a technician."

If technicians feel low on the totem pole, patients feel even lower. A patient who questions the need for an x-ray examination, or is concerned about the haphazard manner in which it is being conducted, runs the risk of being treated with ridicule and disdain by physicians, radiologists, and technicians. The patient is clearly at the bottom of the "pecking order." One woman who questioned the radiation hazards was told, "You're a sissy" by her doctor. When another woman questioned an x-ray department nurse about x-ray exposure, the nurse referred her to the x-ray technician with the remark, "This lady is afraid of x rays. Tell her about them." Sometimes physicians and dentists react to patient questioning with open hostility. For example, a dentist described earlier told his questioning patient that "I don't need this kind of hassle," and he said she'd have to find another dentist.

Quite often nobody takes the time to inform patients about the time, physical discomfort, and expenses which will be involved in a particular examination. This is often considered either someone else's responsibility or no one's responsibility.

There are no easy ways to improve communication among patients and personnel in the diagnostic x-ray system because each individual plays such a specialized role and is part of a hierarchy. However, there is a clear need to develop a better understanding of how the diagnostic x-ray system as it currently exists affects the people who participate in it. For example, what are the rewards or incentives for x-ray personnel who do a good job? What are the punishments for doing an inadequate job? Is the system too much like an industrial assembly line, in which a sense of community effort and pride is thwarted by an environment which makes communication and recognition for a good contribution difficult?

A technician who wrote to me said that he was taught to use shielding when he received his training, but no lead shields were available in his x-ray department and he was afraid to ask for them. Another technician who was also trained to use shielding when appropriate said the radiologist in charge of her facility didn't allow the technicians to take the extra time to use shields.

Many technicians who wrote complained that the workloads in their departments were so heavy they didn't have time to set exposures carefully or to properly adjust the size of the x-ray beam. These technicians clearly didn't have time to communicate effectively with their patients either.

All the people working within the diagnostic x-ray system should cooperate more closely in the mutual sharing of responsibility for the well-being of the patients. These people need to give intelligent patients a more significant role in taking responsibility for their own welfare, because the patient has a right to know what is happening and why. The present diagnostic radiation system is structured so that the patient is the only one who communicates with most of the parts. And he or she could potentially serve as an integrative factor which draws the parts of the whole together.

Patients can also work through consumer groups to encourage better management of x-ray technology. More effective management includes improving communication among physicians who order x rays, radiologists, x-ray technicians, and patients. It also includes providing the right incentives for good work, improving the training and credentialing of all those who prescribe and use x rays, and the reduction of work loads in frantically busy x-ray departments. The result should be fewer unnecessary x rays, fewer retakes, high quality x-ray images, and a reduction in average radiation exposure for each type of examination.

CHAPTER 10
WHAT YOU CAN DO
ABOUT THE PROBLEM

Consumers must play a more active role in their own health care. They must challenge and question their doctor. They must realize that they— and not the physician—have the final say about decisions which concern their health.

Arthur Levin, M.D.
Talk Back to Your Doctor

I feel a little bit of knowledge and a lot of stubbornness can go a long way in cutting down on x-ray radiation if you're willing to be thought of as a crank.

A Patient

Developing New Attitudes Toward Health

Current attitudes among consumers and professionals toward the role of x rays in medicine and dentistry are shaped by attitudes towards health and health care in general. If x rays are to be used more wisely, these attitudes must change.

Modern medicine tends to be "crisis oriented" with an emphasis on the intensive treatment of catastrophic illnesses. Our fascination with medical crises is exemplified by a book entitled *A Coronary Event,* in which a heart attack victim and his physician describe their personal reactions to a heart attack. One reviewer described the book as "in-dispensable . . . for every middle-aged man preparing for his first coronary." The assumption that every middle-aged man should be undertaking emotional preparation for his first heart attack seems in-credible. The assumption can generate a great deal of anxiety among middle-aged men and their families, and it encourages psychological dependence on the health care system.

Criticism about the crisis orientation of medicine as well as our anxieties about health have led to the promotion by the health care system of "preventive" medical procedures, such as periodic checkups, early treatment, mass screening, immunizations, and similar public health measures. The word "preventive" seems misused in this context, since "preventive" medicine merely allows for the early detection and treat-ment of an existing illness rather than its actual prevention. Genuine "preventive" health maintenance should involve the acquisition of per-sonal habits which would reduce the actual incidence of diseases. The idea of preventive medicine as a guardian of good health is typified by the diagnostic clinic operated at the luxurious Greenbrier Hotel in West Virginia.[1] At the Greenbrier Clinic business executives, local clients, and celebrities are examined from head to toe by seven internists. This team of internists is backed by two radiologists, seven x-ray tech-nicians, eight nurses, and six laboratory experts. The x-ray department has six diagnostic rooms, each of which contain fluoroscopic units equipped with closed circuit television screens. One room also contains special equipment to perform mammograms. The Greenbrier phi-losophy was described in a *Parade* magazine article—"Check your body while you're well, *not* when you're sick, and you'll live to a ripe old age."[2] The author of the article asserted that ". . . the speedy detection of almost all perilous symptoms makes them remediable or controllable and, . . . the Clinic will suggest appropriate hospitals, surgeons, specialists, or psychiatrists." Much to his relief, the author was given a "satisfactory report card" by his internist and he intends to return for his next annual checkup.

The myth that the annual checkup and other preventive tests when coupled with appropriate treatments will insure good health is not often discouraged by physicians. For example, according to an article in *Medical Economics:*

Still another criterion for testing is your patient's state of mind. Some patients feel cheated unless you do something for them—a pill, a test. An internist says, 'If a patient seems worried about being anemic, it doesn't matter if she looks pink to me—I take a blood count.' A surgeon comments simply, 'If a patient asks, I give.' Allaying a patient's anxiety by providing objective proof that nothing is wrong is a legitimate reason for ordering a test. And such responsiveness helps build the rapport that staves off suits.[3]

Thus, many physicians feel that anxious patients expect elaborate testing, and they are responding to these expectations.

We have already seen that many, but not all, of the diagnostic x rays associated with crisis-oriented emergency care and routine examinations are of doubtful medical benefit. The approach to health maintenance practiced by our modern technological health care system has resulted in a horrendous waste of x rays and other elaborate, expensive, and potentially hazardous diagnostic procedures.

Fortunately, there is a growing recognition that health maintenance cannot be accomplished with preventive medicine alone. Doctors Donald Vickery and James Fries, in their medical guide entitled *Take Care of Yourself,*[4] strongly advise patients who want to insure good health to change their personal habits. They cite as major causes of modern diseases such habits as drug and alcohol use, smoking, inactivity, and poor diet. These authors assert that:

If we could eliminate all unhealthy habits. . . . Lung cancer and emphysema would almost completely disappear, death due to all cancers would decrease by 25 percent, cirrhosis of the liver would become a rare disease; peptic ulcers, gastritis, and esophagitis (inflammation of the stomach and esophagus) would decrease in frequency; massive upper GI hemorrhage (bleeding from the stomach) would be unusual; pancreatitis (inflammation of the pancreas) would be rare; elevated blood pressure would be less common; atherosclerosis (hardening of the arteries) would decrease in frequency; and accidental injuries would become less frequent. Without the help of the patient, medicine can make no such promises.[5]

In reality, we the consumers of medical care bear the ultimate responsibility for our health and its maintenance. It makes no sense to relinquish this responsibility to physicians and dentists, filing a malpractice claim when our expectations for the outcome of treatments are not met. This responsibility begins with individual health maintenance—rather than with preventive medicine. In many cases, it involves getting rid of unhealthy personal habits which are difficult to change. The payoffs are not instantaneous and they may take years to fully realize. Even then, such changes in lifestyle are no ironclad guarantee against illness, for serious illnesses do occur even in people who have successfully avoided unhealthy personal habits. There are clearly times when appropriate medical or dental care can be of great benefit. But as guardians of our own health, we have a responsibility to learn much more about the health care system, the nature of various illnesses which affect us, and currently accepted methods of treatment. This enables us to strive for a more creative partnership with those who treat us. We also have no obligation to patronize and pay for the services of physicians and dentists who reject the idea of this active partnership, or those who do not seem competent.

Some patients have felt so victimized by their medical care that they have become unwilling to subject themselves to any type of medical treatment. For example, in 1975, a machinist from New Jersey became disabled with chronic pain in his lower back and neck. After visiting many specialists and receiving literally hundreds of x-ray exposures, many of them needless repeats, his problem was not diagnosed. His wife contacted me in 1976 and reported that he was still in pain, unable to work, and that he had run out of medical insurance and disability pay. She wrote:

> To say we are disgusted and disillusioned is putting it mildly. We have given up on the entire medical profession. As far as we are concerned, the American Medical Association, doctors, hospitals, and most other 'medical' counterparts are just interested in making more and more dollars. . . . Well, our coverage is about to run out . . . so we cannot afford any more doctors, so-called specialists, $100-a-day hospital bills, and radiological ripoffs. . . .

> God knows what will happen next. We do know one thing though —*no more x rays come what may.*

> Needless to say we are very bitter towards the entire 'unhealthy' medical structure in this country.

124

This couple's experiences with modern medical care have been considerably more negative than those of most people. Nevertheless, some of the expense and discomfort to which the machinist subjected himself might have been avoided if he and his wife had learned more about lower back ailments, as well as about x rays and other diagnostic tests used to detect lower back conditions.

Learning More About Medicine and Dentistry

My first strategy for dealing with a technologically oriented medical system would be to minimize my own dependence on it by learning about and developing habits to help me maintain my health. Part of educating myself about health care involves learning about ways to treat some of my own ailments without drugs or other specific medical and dental treatments when such treatment is feasible. My second strategy would be to learn as much as possible about those aspects of medicine and dentistry which affect me directly. By learning about the successes and limitations of the health care system, I can become a more informed consumer of health care services. For example, since I have learned more about diagnostic x rays, I feel less dependent on the health care system, less victimized by it.

One of the fundamental problems of conscientious physicians and dentists is keeping up with the explosion of knowledge in their fields. A general practitioner or dentist may be responsible for simultaneously directing programs of diagnosis and treatment for more than a hundred different patients, each patient with a unique set of health problems. Although practitioners ought to keep up with the medical literature, this is extremely difficult for them to do because the rate at which information about health care is produced is overwhelming. Therefore, a knowledgeable patient can really be an asset to an overworked physician or dentist.

I suggest that patients who are told or who suspect that they have a certain illness find out as much as possible about various approaches to its diagnosis and treatment. Several basic medical references are quite helpful in providing background information. For those desiring a brief description of more conventional medical approaches to the diagnosis and treatment, the *Merck Manual of Diagnosis and Therapy* is an excellent and relatively inexpensive home reference. The 1901 edition of the classic *Gray's Anatomy* has been reprinted in paperback by Running Press. It contains a wealth of detailed anatomical drawings and terminology. A good medical dictionary is indispensable to anyone wanting to decipher the thousands of specialized medical words and abbreviations used by physicians and dentists. There are several excellent dictionaries. The one I use is Taber's *Cyclopedic Medical Dictionary.*

An obvious next step is to look for books and magazine articles written for laymen on the suspected illness. The local library is a good source for this information.

Several excellent laymen's guides to medical care are now available. Two that I consider outstanding are *Talk Back to Your Doctor: How to Demand (and Recognize) High Quality Health Care* by Arthur Levin, M.D., and *Take Care of Yourself: A Consumer's Guide to Medical Care* by Donald Vickery, M.D., and James Fries, M.D. The latter guide lists a number of common complaints and gives advice about when these complaints require professional medical attention and when they are self-limiting or might respond to home treatment.

Most libraries subscribe to the *Reader's Guide to Periodical Literature.* Current magazine articles from popular magazines are listed by subject in this guide. Another source is the *Index Medicus;* when I first wanted to learn more about diagnostic x rays, Dr. Sidney Wolfe from the Health Research Group suggested that I look for appropriate articles in this publication. *Index Medicus* is a kind of reader's guide to medical periodicals and is an excellent resource for anyone wanting to learn about current findings in medicine and dentistry. This index is available in medical school, university, college, and hospital libraries. Articles in an area of interest can be found readily in a subject list which covers thousands of topics, such as diagnostic x rays, breast cancer, tuberculosis, and so forth. In addition, most medical schools and university libraries subscribe to the leading medical journals containing the articles listed in *Index Medicus.* I have no formal training in medicine or dentistry, but I have found myself researching all sorts of medical topics this way. Although some of the articles are filled with obscure medical jargon, most of them are surprisingly readable. A good medical dictionary and a careful study of the summaries contained in the articles of interest will help you understand even those articles which are exceptionally technical.

A number of neighborhood clinics and womens' groups throughout the United States have organized self-help groups for people who want to teach each other about health care and how their bodies function. Organizing or joining such a group might be an excellent method of learning more about common illnesses which affect people.

Reading articles about medical and dental research or participating in a self-help group will not qualify you to be your own physician or dentist. However, it may provide you with enough background to discuss your diagnosis and the course of your treatment more intelligently with your practitioner. In any event, it should enable you to take a more active role in your health care.

Handling Ridicule

Many physicians and doctors feel threatened by the notion that patients should play a more active role in their own health care. One radiologist who read my *Consumer's Guide to Medical and Dental X rays* had the following reaction to some of my recommendations:

> As a matter of general philosophy I resent the intrusion of consumerism into the practice of medicine, especially where the physician's judgment is called into question. I am all for upholding high standards of medical practice, but intruding another third party (in this case the consumer advocate) between the doctor and his patient will not insure the maintenance of high standards, and in fact may operate in just the opposite fashion.

I sincerely hope that my role as a "consumer advocate" is not seen by most practitioners as that of an outside intruder who helps in reducing the standards of medical and dental care, because it is my conviction that informed, responsive patients who read this book will be in a better position to improve the standards of their care.

A physician who heard me speak on a radio interview about the problems of x-ray overuse sent me a very short and pointed letter in which he said, "I wonder if you might make the same stupid statements as you were enticed into making via radio today . . . after you have assumed some maturity?"

I would not remain under the care of any physician who was as openly hostile as this one. On the other hand, I would not wish to dictate how my medical or dental care ought to proceed on the basis of reading a few articles. (I have had a fairly good response to my attempts to share copies of articles related to my symptoms with professionals and then discuss them.)

Physicians, dentists, and x-ray technicians who react defensively to their patients' questioning often treat these patients with ridicule. Some of the patients who wrote to me reacted to this ridicule in various ways. One woman stated:

> Ridicule is the norm when one questions the need for an x ray. The personnel try to make you feel ignorant or guilty. Most people go through the x rays just to avoid the hassle. I have solved the 'dental' problem by faithful use of nutritious foods and mineral supplements. My teeth are clean and tight, fillings are solid, and the health of the gums is excellent—much better than when I went 'every six months'—a real rip-off, to be sure.

A maternity patient wrote:

> I questioned the use of x rays on new mothers . . . , but I was ridiculed and given the old story about the woman whose tumor or lump was discovered after taking one of these valuable x rays. I didn't buy the story, but I did give in to get them off my back.

Another woman made a special appointment with her dentist because he repeatedly ignored her requests to relinquish the dental x rays to her. During the appointment she described his manner as "firm, annoyed, and patronizing." Although she left "feeling very put down," the dentist mailed the x rays to her shortly after the appointment.

A fourth woman described a situation which was not as productive:

> Just a few weeks ago I had a consultation with my family physician. Knowing from the past how eagerly he orders x rays, I asked him to limit the x rays to those which are absolutely necessary, and wherever possible resort first to other tests that would serve the purpose. My doctor became very angry. He told me that either I will do what he wants me to do or he will not examine me at all. Well, he did not examine me and I left his office in tears. I have never felt so defeated in my life.

A mother who witnessed a dentist and his technician taking six x-ray exposures before a readable one was obtained asked that neither of her children be x-rayed in the future. She said "they complied, but looked at me as though I was an interfering and overly protective mother." When this mother moved to a new community she again asked that no x rays be used on her children. Her new dentist asked her to sign a release form and said, "there is no more radiation in an x ray than there is in ten minutes in the sun."

A nursing mother who found out she was being taken for a routine hospital chest x ray asked an attendant to return her to her room. He became angry but complied with the request. She reported: "I was nursing the baby and extremely happy that I had enough spunk to refuse the x ray. I'm certain from the reaction that no one else did."

Clearly some patients find themselves submitting to x rays they don't think are needed in order to avoid embarrassing confrontations. Others have simply refused to submit to certain x-ray examinations or have found practitioners who feel less threatened by their patients' requests to reduce unnecessary x-ray exposures.

It is possible to find sympathetic practitioners, and one couple who wrote to me described a positive experience with their dentist. This couple was concerned about the possible effects of regular dental x rays on their two young children. The father had phoned his family dentist to request that dental x-ray examinations not be performed on his children every six months. The dentist listened "very respectfully" to his views. Then the dentist said he would comply with the father's wishes if he would send a letter which could be filed in the dentist's office. The couple composed the following letter:

> Pursuant to our phone conversation today I am writing this letter to explain our reasons for not wanting our children to be x-rayed every six months.
>
> Even though most dentists find the x-ray machine an important diagnostic tool in locating problems invisible to clinical examination, we do not feel that exposure to radiation every six months or even every year is necessary.
>
> My wife and I (and many others, including radiologists and some

129

physicians) feel that the effects of radiation exposure are cumulative. The damaging effects cannot be detected for at least 15 to 20 years afterwards. Continuous and cumulative exposure to radiation has been known to damage tissues and organs. It has been cited as a cause of cancer and other disorders including genetic damage.

We think that the use of the x-ray machine, as a diagnostic tool, should only be used in an emergency or where difficult problems are suspected.

We are indeed grateful that you expressed respect for our views and will continue your good work in treating our family professionally to the best of your ability without the use of x ray.

This letter, we hope, will serve to release you from any sense of frustration you might feel in not being able (at our request) to use your full diagnostic skills in the treatment of our children's dental needs.

If, at any time in the near future, we learn that x-ray diagnostic practice has improved and that dentists and doctors exchange x-ray data to prevent duplication, then we will reconsider. Until then, take good care of our children's dental health with the required restriction.

This letter not only spelled out the couple's reasons for their request, but it indicated that these patients were willing to take responsibility for their decision. The letter also indicated that the couple had a sincere respect for the contribution which the dentist was making to the family health. When the father wrote to me, he suggested that others might want to write letters "releasing" their dentists from the obligation of "employing what they consider a very important diagnostic tool." This suggestion is a good one, and similar communications with dentists and physicians could improve doctor/patient relationships and at the same time reduce unnecessary x-ray expense.

Selig Greenberg noted in his book about hospitals, *The Quality of Mercy,* that "medicine is the only big business in which the ultimate consumer not only lacks the requisite knowledge for making a rational choice but has almost no control over what he buys." [6] Reducing unnecessary x rays will not be accomplished unless consumers themselves insist on and work towards changes in the health care system. Consumers should be serving on governing boards and committees for hospitals and clinics. They should serve on medical or dental insurance company boards. Concerned individuals should join or organize consumer action groups which can work to change policies or legislation related to the conduct of x-ray examinations.

Legislative changes would be beneficial in the licensing of radiologic technicians and on the setting of exposure limits for diagnostic x-ray procedures. Policy changes are needed with regard to abdominal x rays of pregnant women, mobile unit chest x rays and routine hospital chest x rays.

On an individual level, the measures which will help reduce x-ray exposure are basically the same measures which will enable us to get better health care in general. Here are the things we can do:

1. Learn how to reduce the chances of ill health by avoiding unhealthy habits.

2. Learn about the limitations of modern health care. Which diagnostic tests are reliable and have a high yield? Which treatments seem most reliable and safe? Which illnesses require consultation and self-treatment and which are self-limiting, that is, cured without medical attention.

3. Learn not to expect medical miracles of physicians and dentists in dealing with every complaint. Instead, try to learn something about your own illness or complaint so that you can develop reasonable expectations of treatment, compatible with current knowledge in the field.

4. Learn some of the ways to treat the simple conditions which do not require the care of a physician or dentist.

5. When consulting a physician or dentist, select one who appears competent, who understands your approach to health care, and who is willing to share the responsibility for it with you. (Some people are finding it helpful to pay for an initial interview appointment before choosing a practitioner or dentist.)

6. When undergoing diagnosis or treatment for a serious condition, learn as much as possible about the condition and about the accepted treatments for it.

131

Living with complex technologies has its joys and its sorrows. A sophisticated diagnostic x-ray examination leading to the successful treatment of a serious illness can be a source of great human comfort. On the other hand, coping with the complex health care system which accompanies this health care technology can be frustrating, expensive, and dangerous. Those of us who choose to indulge in the benefits of such a system must learn about its risks. Coping adequately with modern health care demands that we learn much more about its strengths and weaknesses. In so doing we are acting in our own interests. In short, all of us must accept full responsibility for our own health.

CHAPTER 11
MORE ABOUT
DIAGNOSTIC X RAYS
AND THEIR
ALTERNATIVES

Since their discovery in the last century, x rays have been used repeatedly for the detection of dense objects such as broken bones, metallic objects, and gall or kidney stones. In the years between 1895 and the present, x-ray diagnostics has developed into an increasingly fascinating and sophisticated science.

What Are X rays?

To scientists, x rays, like visible light and radio waves, are bundles of energy which move at the speed of light and which have electric and magnetic fields associated with them. Because x rays are more energetic than light, they can penetrate material and collide with some of the atoms of which the material is composed. These collisions can result in the separation of an electron from an atom, or ionization, and thus, x rays like gamma, beta, and alpha rays are classified as ionizing radiation. When ionization occurs in the atoms of living tissue, it can produce biological damage.

If a section of a human body is placed in front of a beam of x rays, some of the x rays will pass through while others will be absorbed or scattered inside the body. Denser materials such as bone absorb x rays more readily than surrounding tissues. When the x-ray beam exposes photographic film or a fluoroscopic screen, the shadows of the denser parts of the body appear on the developed film or screen.

Techniques in X-ray Diagnostics

The scope of x-ray diagnosis can be expanded greatly when substances known as contrast media are used. Air was the earliest and easiest contrast medium to be used. A deep breath held during a chest x ray fills the lungs with air. X rays pass through the air so readily that the tissues surrounding the lungs show up easily. Thus, any areas where the lungs do not readily fill with air show up as cloudy spots on an x-ray film.

While x rays pass easily through air, they are absorbed by dense substances like barium or iodine, known as radiopaque materials. These can be used as contrast media if introduced into the body before an x-ray examination is conducted. Radiopaque materials can be swallowed, injected into the general bloodstream, or introduced locally using a relatively new technique known as catheterization. In this technique, a plastic tube or catheter is inserted into a vein or artery and guided through the blood vessel to the location of interest under x-ray observation. A contrast medium can then be passed through the catheter. Taking x-ray pictures of veins and arteries is called angiography. The blood supply system of almost any part of the body can be studied using this technique. Barium and air are often used as the contrast media in gastrointestinal examinations. The patient may swallow a chilled, slightly sweetened barium drink, or be given a barium or air enema.

Early x-ray scientists developed methods of projecting x rays which pass through patients onto fluorescent screens that give off visible light when excited by x rays. This method, known as fluoroscopy, has enabled physicians and radiologists to observe the motion of organs or materials inside the body by viewing a patient continuously. A radiologist can watch a barium liquid pass through a patient's esophagus to his stomach.

Early fluoroscopes gave off a dim green light when examinations were performed in a darkened room. This required examiners to spend about 10 minutes of precious time adapting their eyes to the dark before concentrating on a dimly lit screen. Mass-survey chest examinations are often conducted with a photofluoroscopic technique, in which a fluoro-scopic image is photographed, and the photographic image is stored on a small film. In this case, fluoroscopy is used to aid in reducing the size of the x-ray film rather than to observe a process in motion. Although fluoroscopic and photofluoroscopic techniques sometimes provide important information which cannot be provided any other way, they take longer to perform, and they often expose patients to more radiation than conventional x-ray examinations recorded on film.

Modern fluoroscopes, however, usually involve significantly less x-ray exposure than the older models because of the recent development of electronic image intensification systems. These systems amplify the light from the fluoroscopic screen and then provide a brighter image on a TV monitor. Newer fluoroscopes are also equipped with foot pedals that turn the x-ray tube on and off, as well as with switches that allow an examiner to record interesting views on x-ray film for later examination. These new developments provide better information, and if they are used properly can reduce patient exposure by a factor of 100 or more.

The typical exposure to x rays has also been greatly reduced during conventional x-ray examinations by the development of new high speed x-ray films and improved development techniques. Phosphorescent screens which envelop the x-ray film are now used for many types of examinations. These screens allow an image to be formed on the film with a much lower x-ray dose. Improved techniques and equipment have increased the usefulness of diagnostic x rays far beyond the early days when x-ray examiners were limited to the observation of "bullets, bones, and gallstones."

Even newer methods which will further expand the capability of x rays as a diagnostic tool are on the horizon. Many of the recent and proposed developments will probably decrease the exposure of patients to potentially harmful radiation in a given examination. For example, new rare-earth-type intensifying screens which are over two times more sensitive to x rays than conventional screens are now under development. These new screens should make it possible in the near future to cut patient exposure in half without changing the quality of the x-ray image. Also, certain new types of examinations such as those using CAT-scanning (Computerized Axial Tomography) may expose patients to less radiation than conventional x rays. However, as the capabilities of x-ray diagnoses expand, so do the number of different examinations to which a typical individual is likely to be subjected. In the United States the frequency of both medical and dental x-ray examinations increased significantly between 1964 and 1970.[1]

How X rays are Produced

X rays are usually generated in a glass tube from which all the air has been removed. The x rays are generated when a current of electrons (negatively charged atomic particles), measured in milliamperes, is stopped suddenly by a metal target. The speed at which the electrons hit the target depends on the voltage which is placed across the tube. Thus, increasing the voltage across the tube increases the energy of the electron current and produces a more energetic and penetrating beam of x rays. Some of the less energetic x rays leaving the tube are filtered out and absorbed by the glass walls of the x-ray tube.

During the time that an x-ray machine is on, the voltage which produces the x rays may vary in a regular periodic manner between zero and a maximum value. The maximum voltage is often referred to as kilovolts peak (1 kilovolt = 1000 volts) or kVp. Diagnostic x rays are usually produced at voltages which range from 20 kVp for low energy x rays to 150 kVp for high energy x rays. At a given kVp the x-ray beam which is produced has a distribution of x-ray energies between zero and a maximum value which is the same as the peak voltage impressed across the tube. The distribution of the x-ray energies in the beam determines the quality or penetrating power of the beam. The quantity of the beam, or the total number of x-ray particles in the main beam emerging each second, depends on both the peak x-ray tube voltage and the current, or number of electrons stopped by the metal target each second. The total number of x rays striking a patient being examined is proportional to the electron milliamperage and the number of seconds (s) that the beam is on. Although the beam quantity also depends on the tube voltage, or kVp, the total x-ray quantity is often measured by machine operators in milliAmpere-seconds or mAs.

Some of the energy contained in the electron current within this x-ray tube is used to heat the target plate which stops the electrons. To prevent over-heating and damaging the target plate, many x-ray tubes contain a larger circular plate which rotates rapidly while the electron current is on, distributing the heat over a larger area of metal. The whirring sound of this rotating plate, or anode, is what one usually hears when an x-ray tube is on.

Since many of the lower energy x rays in a beam do not penetrate tissue very well, they do not pass through enough tissue to contribute to an x-ray picture. Many of these low energy x rays, however, are absorbed by the outer layers of skin and could cause needless biological damage. For all but a few types of diagnostic examinations the useless low energy x rays can be easily filtered out of the beam by placing a sheet of aluminum which is 2 to 3 millimeters thick (about $\frac{1}{10}$ of an inch) between the x-ray tube and the patient. This filter improves the "quality" of the x-ray beam produced by the tube.

An operator of a diagnostic x-ray machine can determine the beam quantity by adjusting dials to give a desired peak x-ray tube voltage (kVp), electron current milliamperage (mA), and time in seconds (s) for the voltage to be impressed on the tube. The beam quality varies with the voltage and current settings, as well as with the adjustment of the thickness of the metal filter. An operator will usually choose a different combination of kVp, mAs, and filtration depending on the type of examination and size of the patient.

New Alternatives to X rays: Ultrasound, Thermography, and Nuclear Scanning

There are several relatively new technologies which enable radiologists to obtain pictorial images of body structures. Methods known as ultrasonics, thermography, and nuclear scans are beginning to replace x-ray examinations in some cases.

The use of ultrasound (sound waves of frequencies above the range of human hearing) has gained rapid acceptance as a diagnostic tool in recent years. Since 1970 the sales of ultrasonic equipment have been increasing rapidly, and it is estimated that they will rival that of x-ray equipment in the 1980s.[2]

In ultrasonics, information about the internal structure of the body is obtained by recording the way sound waves are reflected from different parts of the body.

The same device which emits short pulses of high frequency sound serves as a detector for the reflected pulses. If the sound source is moved continuously, an area of the body can be scanned. These reflected pulses can be transformed into an electrical signal which can activate a television screen and produce an image of the area being scanned.

Although ultrasonic scans can be used as a substitute for x rays for certain kinds of examinations, they yield different types of information. The boundaries between different types of soft tissue can be highlighted with ultrasonics and characteristics of blood flow can be studied. However, x rays are still better for the visualization of anatomic structures such as bones.

So far, the most popular applications of medical ultrasonics lie in the diagnosis of heart abnormalities, eye conditions, arterial blood flow, and brain problems caused by tumors, cysts, or hemorrhages.

The use of ultrasonics in diagnosing problems in pregnancy is also becoming more common. Ultrasound has been used extensively on pregnant women as a replacement for pelvimetry and other fetal examinations. The amniotic fluid surrounding the fetus is an excellent transmitter of ultrasound, while the fetus serves as a reflector. This allows physicians to use ultrasound to view an image of a developing child on a television screen. Ultrasonic scanning of the uterus can be used to detect pregnancy as early as the sixth week, to determine the number and size of fetuses, and to determine the fetal position. Ultrasound has also been used to monitor the position of a needle in cases where it is desirable to withdraw some amniotic fluid for further study and to detect large fetal deformities. A major reason for the popularity of ultrasonic scans in pregnancy is the assumption that ultrasound, unlike x rays, does not cause fetal damage.

Even though no studies have yet revealed any biological damage in humans from diagnostic exposure to ultrasound, nevertheless, it is a form of energy being transmitted through tissues, and it may, like x rays, eventually be proven harmful. When x-ray diagnosis was first used practitioners were not aware of the subtle, long-term genetic and cancer-producing potential of low intensity x-ray beams. Many of the effects of low level radiation are not apparent until 15 or more years after one has been exposed to them. This may also prove true for ultrasound, but we are hoping that it will not, and that ultrasonic diagnostics will at least provide a risk-free alternative to x rays for certain procedures—especially those now exposing pregnant women to abdominal x rays.

The full potential of ultrasonics has not yet been realized. Medical researchers are exploring a host of new applications for the technique, and engineers and physicists are designing and testing new equipment. The first generation of ultrasound scanners are difficult to use, and the quality of the image is highly dependent on the skill of the machine operator. Therefore, new developments will be needed if ultrasound is to be used effectively.

Another new development, thermography, measures the infrared radiation pattern resulting from temperature differences on skin surfaces. Commercially available thermographic systems consist of an infrared detector, camera, and display unit. The infrared detector scans the area of interest. Because the temperature patterns are already present and no energy is transmitted through the body, thermography is absolutely safe. Thermography is widely used in examinations of the breast and other organs, but like ultrasound it does not always yield the same information as x rays. Thermography is used along with mammography in the breast cancer detection clinics described in Chapter 6. Unfortunately, thermographic scans are not yet as reliable as mammography for the detection of very small breast tumors. If they were, there would be no need to use potentially harmful x rays.

Nuclear scans have been commonly used for the detection of tumors since the early 1950s. They are also used to study the amount of liquids located in organs, the rate of flow of liquids through organs or membranes, and the changes in internal organs during a period of time.

A nuclear scanning procedure begins with the introduction of a radioactive substance into the body. (A radioactive substance is one whose atoms give off radiation such as beta particles or gamma rays, which are similar to x rays.) The way in which a radioactive material passes through the body depends on the properties of the atom or molecule with which it is associated. Thus, after a radioactive material (or tracer) is introduced into the body its location can be determined by a detector which scans over the body and records how many beta or gamma particles are being emitted at each location. The number of particles at each scanning location can be transformed into a spot on a photographic film or a television camera where the brightness of the dot will be proportional to the number of particles detected. If a tumor or abnormal growth exists in an organ, more or fewer molecules of a particular radioactive chemical may tend to accumulate there than in the surrounding tissue, and the tumor would show up as a bright or dim spot on the scan picture. Since there are many different types of radioactive elements, almost any kind of molecule can be "tagged" with a radioactive atom. Thus, physicians who specialize in the field of nuclear medicine have many different tracer substances which can be used for a variety of nuclear scanning procedures.

Since some of the radiation given off by the tracer substance is absorbed by the body, the risks associated with nuclear scans are similar to those associated with diagnostic x rays. In fact, some scans deliver a considerable radiation dose to the organ being investigated, such as the thyroid gland. Therefore, nuclear scans cannot be thought of as safe alternatives to x rays. Nevertheless, when a nuclear scan is needed, it can provide useful information, and nuclear scanning will undoubtedly remain an important diagnostic tool.

CHAPTER 12
HOW X RAYS
AFFECT PEOPLE

Determining Effects on People

The potential for danger from x rays and other forms of radiation has often gone unnoticed in the past because the damaging radiation cannot be seen or felt. Except for situations in which people are exposed to enormous amounts of radiation, signs of damage are not noticeable until some time after exposure, perhaps a number of years.

Three units are commonly used in discussions of the potential biological effects of x rays and other types of radiation: the roentgen, the rad, and the rem. The roentgen is an approximate measure of the radiant energy to which a person or object is exposed, while the rad and rem provide a measure of the amount of energy actually absorbed by an exposed object.

Although on a technical level each unit is defined differently, for the purpose of the general discussion in this book, these units can be used interchangeably. One roentgen, rad, or rem represents a fairly large amount of radiation when compared to the average natural background dose received annually. For example, the average whole-body radiation dose received from natural sources (background radiation) in the United States is about 0.084 rems per year.[1] Often millirad (mrad), millirem (mrem), or milliroentgen (mr) are used instead. The prefix milli means 1/1000, so that 1 rem = 1000 mrem. Expressed in terms of the smaller units, the average whole-body dose in the U.S. due to background radiation is 84 millirems per year. (See Appendix A for a more detailed discussion of these units.)

It is estimated that the average resident of the United States receives an abdominal dose of 73 mrem per year,[2] primarily from medical x rays. Thus, the average abdominal dose received from diagnostic x rays is just slightly less than the abdominal dose received from natural background radiation in the United States.

Effects of Moderate and High Doses of Radiation

Experiments with animals and observations of people exposed to much larger doses of radiation than those associated with diagnostic x rays have provided valuable information about effects of radiation on living organisms.[3] The known effects are of two kinds. One set of effects directly influences the health of the exposed individual. These effects are known as *somatic effects*. Another set of effects which influence the health of the off-spring of the exposed individual are known as *genetic effects*, and are produced only when the reproductive organs (ovaries and testes) are exposed to radiation.

Somatic Effects

A study of American radiologists exposed to medical x rays in their occupations and two control groups consisting of other physicians indicated that on the average, radiologists in the United States have a higher incidence of cancer than their medical co-workers.[4]

The tragic exposure of a large number of Japanese people to the very high levels of radiation from the atomic bomb blasts in Hiroshima and Nagasaki in 1945 has added greatly to our understanding of the harmful effects of radiation.[5] When very high doses of radiation are absorbed rapidly (when more than 50 rems are received at a rate faster than 1 rem per minute), like those received by some Hiroshima victims, radiation sickness and in some cases death occurs within a few days or weeks after exposure. Other Hiroshima survivors who received somewhat less radiation and lived for a period of years after the atomic explosion had significantly more cataracts, cancers, and leukemia than an equivalent unexposed population. Pregnant women who received large doses of radiation and who survived the holocaust bore children who had an abnormal number of birth defects and a greater chance of developing cancer or leukemia. The number of miscarriages and still-births for this group of pregnant women was also significantly greater than average. Young children exposed to radiation suffered from many more health problems in subsequent years than adults receiving the same doses, although it was found that survivors of all ages who absorbed radiation had a greater risk of developing cancer or leukemia, or giving birth to deformed children in later years.

These studies of atomic bomb survivors indicate that when large amounts of radiation are absorbed rapidly by an individual, the risk of subsequently developing cancer seems to be directly proportional to the radiation dose received. Thus a victim who absorbs twice as much radiation runs twice the risk of developing cancer in later years. This same type of proportional relationship between high radiation doses and chances for developing leukemia and certain other long term somatic effects has also been established for animals tested in laboratories, and for children who were exposed to diagnostic radiation before they were born,[6] as well as for British patients who received large doses of x rays while being treated for a spinal disease known as ankylosing spondylitis.[7]

Genetic Effects

There is no entirely convincing data available on the ways in which radiation actually damages the genes and reproductive organs of humans. For the most part, we know these organs must be damaged by observing the higher rates of genetic illnesses present in the children of people whose reproductive organs have been exposed to high levels of radiation. However, a tremendous number of experiments have been performed on laboratory animals (such as mice and fruit flies) exposed to known amounts of radiation, and scientists believe the results of these experiments are indicative of the kinds of genetic damage incurred by humans. There is evidence that the reproductive material in human cells is somewhat more susceptible to damage than that in the cells of mice.[8] Thus, it is believed that radiation of the reproductive organs of humans can cause offspring to have a higher rate of mental retardation, cancer, ill health, and birth abnormalities. Except at very high doses, the amount of genetic damage done by radiation appears to be proportional to its dose and unrelated to the rate at which the reproductive

147

organs are exposed. Harmful effects on the reproductive material (chromosomes) in human cells have also been observed directly when the cells are exposed to x rays or other forms of ionizing radiation.[9]

It is felt that laboratory data obtained from irradiating mice provides the most significant information about the possible genetic effects of radiation on humans.

The Effects of Radiation on Living Cells

Cells are the building blocks of living organisms. Some scientists believe that both the genetic and somatic damage associated with radiation exposure can be traced to the way in which radiation affects living cells. In any healthy organism some cells are dying while others are reproducing, replenishing, or increasing in number. Cells which are exposed to radiation often die or lose their ability to reproduce properly. Radiation loses its energy in passing through a cell by breaking apart some of the complex chemical bonds in the cell. As a consequence, the chemical bonds characterizing the structure of the genes in the cell nucleus, which directs its reproduction, can be altered either directly or indirectly by the radiation.

The ability of cells to reproduce accurately is quite important, and it is very probable that the impairment of the reproductive ability of cells in the human body results in increased susceptibility to cancer and other diseases. The process of aging is also believed to be related to the reproductive ability of cells in an organism.

When a sperm or egg cell from the human reproductive system is exposed to ionizing radiation, the arrangement of the basic genetic material, which contains instructions for the construction of offspring, may be altered or destroyed. These rearrangements, which are called mutations, may result in the failure of the cell to contribute to the reproductive process. A more serious consequence, however, is that such a damaged egg or sperm cell may produce offspring with physical, mental, or genetic damage. A number of illnesses and birth defects have now been attributed to genetic damage or mutations of one type or another.

It is well established that in humans the potential for harm from radiation exposure is greater for growing individuals whose cells must undergo rapid division to support growth. The embryonic stage is the most sensitive of all to radiation damage. However, in both adults and children some organs and tissues discard and replenish cells more rapidly than others. Organs and tissues which are more sensitive to radiation exposure than others include the thyroid and the red bone marrow where blood cells are manufactured.

Effects of Low Doses of Radiation

The effects on humans of low doses of radiation (less than 25 rem) similar to those delivered during common diagnostic x-ray examinations are not immediately noticeable. Indeed, since the discovery of x rays in 1895, there has been an ever-increasing awareness of some of the more subtle long term health effects of radiation. This fact is rather graphically illustrated by the way in which permissible doses for individuals working in the presence of radiation have declined over the years (see Fig. 1). Since 1940 the permissible whole-body occupational dose has decreased five-fold. Currently, it is believed that a person who is exposed to x rays during a medical or dental examination may have a

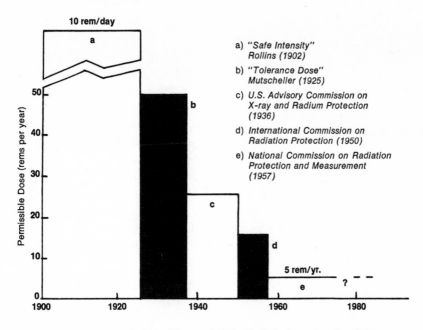

Fig. 1 — Historical trend in permissible whole body occupational dose

slightly greater chance of developing cancer or leukemia in later years. However, even if one of these illnesses develops, it would be impossible to trace its cause back to an x-ray examination. This is due to the fact that individuals are also exposed to other cancer-causing agents such as air pollution, drugs, and natural radiation.

As we noted earlier, at moderately high doses (25 to 50 rem) and dose rates a statistical relationship between radiation exposure and various cancers and leukemias has been observed in both humans and laboratory animals. For low doses, however, it is more difficult to separate the effects of radiation on humans or laboratory animals from other influences in the human or laboratory environment. When conducting experiments on the effects of very low doses of radiation on mammals, scientists need to study a very large group of animals in order to obtain valid results. This makes direct observations on humans quite difficult.

Until recently there was not much evidence that exposure to diagnostic x rays affected human health. One recent study has shown, however, that exposure to diagnostic x rays increases the chances of unborn children developing cancer or leukemia during childhood.[10] Although this study indicates that the probability of developing one of these conditions is in proportion to the radiation dose delivered to the fetus, these results are not conclusive, since it is easy to argue that the relatively poor health of those mothers who had x-ray examinations during pregnancy caused the increased susceptibility of their offspring to childhood cancers. A more recent study, supported by the National Cancer Institute, identifies certain types of children as more susceptible to radiation damage than others. In particular, the study indicates that children with a history of allergies or certain bacterial diseases were up to ten times more likely than other children to develop leukemia after being exposed to diagnostic x rays before birth.[11]

In 1972 a study *linking leukemia with the exposure of adults to diagnostic x rays* was published.[12] In this study the development of excess cases of leukemia in adult males was traced to previous diagnostic x-ray examinations. For some unknown reason the females in the study group did not show a significant increase in leukemia.

Many more sophisticated studies will be necessary before the full story of how diagnostic x rays are affecting humans is unraveled. At present we can only make judicious estimates of how low level x-ray doses affect humans, on the basis of other evidence.

In the laboratory there are quite strong indications that the response of mice to radiation dose is similar to the response of humans.[13] Both genetic and somatic radiation damage to laboratory mice and their reproductive cells have been observed at relatively low doses.

The Linear Hypothesis

Since it is difficult to establish an exact relationship between biological damage and dose for low levels of radiation, many scientists believe that in the absence of concrete evidence it is most prudent to assume that ionizing radiation involves a health risk to the exposed individual in proportion to the dose absorbed by that individual. This is known as the linear and non-threshold hypothesis. This hypothesis assumes there is no threshold below which radiation is harmless. Thus, *under the linear non-threshold hypothesis any amount of radiation absorbed by an individual, no matter how small, involves some risk to the health of that individual and/or his or her potential offspring.*

There is evidence, however, that indicates that certain effects induced by high radiation doses are not caused by lower doses.[14] For example, many Hiroshima victims who absorbed large doses of radiation in their eyes developed cloudy films or cataracts over the eye lenses in later years. Cataracts were also induced in the eyes of laboratory animals exposed to large doses. On the other hand, animals exposed to lower

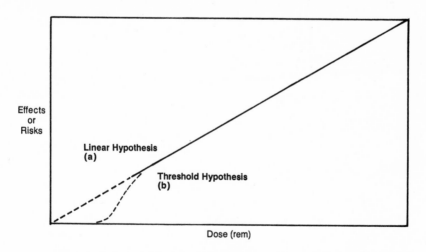

Fig. 2 — Relationship between radiation dose and effect using (a) the Linear Hypothesis, and (b) a Threshold Hypothesis

151

doses of radiation did not develop any cataracts. Cataract induction appears to be a threshold effect in the sense that there seems to be a dose below which radiation does no detectable damage to the lens of the eye. However, many scientists believe that genetic damage can be caused at even the lowest levels of radiation, that there is no threshold for this type of damage.[15] Since the question of whether or not a threshold exists for radiation damage is a matter of controversy, it is interesting to speculate about the problem in terms of the effects of radiation on the cells which make up the human body.

When ionizing radiation, like an x ray, passes through a cell, the cell may be damaged. This damage can either destroy the cell, impair its ability to reproduce, or cause it to reproduce with incorrect genetic information. The effect of high levels of radiation on an organism is serious because many of its cells are damaged and destroyed. At lower doses fewer cells are damaged. In either case, the number of damaged cells is roughly proportional to the amount of radiant energy (or dose) absorbed, no matter how small it is.

If individual cells have mechanisms for repairing radiation damage, then there may be a threshold dose below which radiation causes no permanent damage. However, the 1972 study linking leukemia in adult males to diagnostic x-ray doses suggests that if a threshold dose exists for adult males, it must be less than the doses delivered during some of the common diagnostic x-ray examinations.

In this book, I have assumed that the linear hypothesis is a reasonable working hypothesis for protection purposes. This assumption has also been made by official advisory groups including the International Commission of Radiological Protection (ICRP) and the National Council on Radiation Protection and Measurement (NCRP) in originally establishing the guidelines used for the protection of people working near radiation sources. In addition, scientists representing the prestigious National Academy of Sciences (NAS) gave strong support to the linear hypothesis in its recently published study (known as the BEIR Report) on the effects of low level ionizing radiation.[16]

The American College of Radiology has also taken the position that ". . . the prudent course at the present time is to assume that (radiation) effects may be initiated in a single cell and to assume that there is no lower limit to the radiation dose which might be considered injurious."[17]

CHAPTER 13
RISKS ASSOCIATED WITH X-RAY DIAGNOSIS

The radiation risks associated with an annual exposure to diagnostic x rays span a wide range, from examinations which may be comparable to regular cigarette smoking during a given year, to those which involve negligible risks. Pregnant women, young children, and potential parents are usually subjected to more risks in a given examination than other people. There are additional risks associated with certain diagnostic examinations involving the use of a contrast medium. These risks are caused by the introduction of the medium into the body and not by the x rays.

In general, examinations of organs and tissues in the abdominal regions involve significantly more somatic and genetic exposure than other examinations for several reasons. First, the abdominal region is filled with relatively dense tissues, so more x rays must be used during these examinations. Second, a relatively large film is used, thus exposing a large area of the body to x rays. Third, the reproductive organs are in the abdominal region. Finally, special risks to the unborn child are involved when abdominal x rays are used on pregnant women.

How Risks are Expressed

In addition to radiation exposure, many other typical human activities such as taking drugs, riding in a car, or smoking entail risks. These kinds of risks are classified as somatic, meaning that they are

155

potentially harmful to individuals. Exposure of potential parents to radiation and certain drugs involves genetic risks, which only affect individuals in future generations. Taking both somatic and genetic risks into account when considering diagnostic x rays makes estimation of the risks and benefits more complicated.

Until recently, very little attention was given to obtaining numerical estimates of the risks or potential for harm associated with familiar activities. Even in cases where numbers have been assigned to risks, relatively few people use such information and change their habits or conduct accordingly. A recent exception to this is in the case of cigarette smoking; the link between smoking and lung cancer deaths and other health conditions has motivated many individuals to quit smoking. If we are to begin taking more responsibility for our own health care, we must become concerned about the risks involved in various forms of medical diagnosis and treatment.

Several approaches describe the risks associated with various activities. Some analysts have attempted to devise a suffering scale for health problems that may result from certain activities. For example, using such a scale, a fatal radiation-induced leukemia would be much higher on the suffering scale than a skin cancer which may be painful but which is rarely fatal. Another way to express risk is to place an economic value on induced disabilities or deaths. Such an economic analysis would add medical expenses to the loss of an average individual's income from employment.

The easiest method to use in relating radiation dose to risk is to estimate the number of excess deaths that would be expected if a specified number of people were all exposed to the same hazard. This method has been used to link the probability of developing lung cancer to smoking cigarettes. For example, according to a study conducted by Doctors Hammond and Horn, about 130 additional people out of one million regular pack-a-day smokers will probably die of lung cancer each year.[1]

Risk estimates for various x-ray diagnoses, like that for smoking, must be approximate. Scientists do not know enough about the precise relationship between absorbed dose and risks. Even if they did, the absorbed dose received from specific diagnoses depends on the techniques and equipment used to conduct each individual examination. For this reason, a reasonable way to express the risks involved in each type of x-ray examination is to calculate the risks using good equipment and techniques and those using poor equipment and techniques. We can then assume that the average risk falls somewhere between these two extremes.

What Are the Risks of Diagnostic X rays to Your Health?

Estimating the risk of death from cancer or leukemia for various x-ray examinations is complicated by the fact that it depends on the age and susceptibility of the exposed individual. Also, each examination exposes different organs and tissues in the body to radiation. Areas of the body like the red bone marrow and the thyroid are much more sensitive to damage by radiation than muscle or skin. Recent scientific publications have described the relative likelihood that various critical organs will develop cancer after receiving a given radiation dose.[2] If we know the skin dose (amount of radiation absorbed by the skin), the voltage of the x-ray tube and type of filter used, as well as the organs in the main beam, we can use these figures to compute a cancer producing or "effective dose" for a given x-ray examination.[3]

"Effective doses" are listed in Table 1 (later in this chapter) and Appendix C for some common x-ray examinations conducted under average conditions. A recent report published by the National Academy of Sciences, known as the BEIR Report, estimates using the linear hypothesis (see Chapter 12) that a whole-body dose of one rad or rem received by a million individuals each year causes an average of 80 excess deaths from cancer annually.[4] If the "effective dose" concept is valid, then one rad of effective dose would affect the risk of cancer and leukemia in the same way as one rad of whole-body dose.

The average number of deaths per rad should be considered an upper limit, since it is still possible that discovery of a threshold dose in the future may indicate that some of the lower dose diagnostic x-ray examinations involve little or no risk.

The "effective doses" and thus the cancer risks from different types of x-ray examinations vary widely. As shown in Appendix C, a lower gastrointestinal, or GI, examination conducted under average conditions

157

may involve as much cancer risk as smoking somewhere between five and 20 cigarettes every day for a year. On the other hand, average examinations of the extremities or dental examinations involve less risk than smoking even one cigarette per day. The cancer risk associated with the average U.S. background radiation dose of 84 millirems per year [5] is comparable to smoking about one cigarette a day. Thus, typical examinations span a wide range of maximum risk.

Thus, GI examinations, which deliver the highest "effective dose" of any of the common procedures, may involve risks of fatal cancer comparable to the fatal lung cancer risks associated with regular smoking. (See Tables 1, 2, and 3, later in this chapter, for details.) On the other hand, a hip or thigh examination is about five to 10 times less risky than a gastrointestinal examination, and the "effective dose" of about 100 millirems per year received from hip or thigh examinations is comparable to the average background radiation dose in the United States.

Risks to Future Generations

Genetic risks to potential parents from x rays are difficult to compare directly to the risks of smoking and other activities, but information from the BEIR Report indicates that when the reproductive organs are included in the main beam, risks to future generations are similar to those encountered in submitting to x-ray examinations which are in the high dose somatic category. Table 2 and Appendix C give data on the average male and female reproductive organ doses for various examinations.

Non-Radiation Risks

Some diagnostic x-ray procedures involve the introduction of a contrast medium into the body in order to highlight certain organs or arteries. Contrast films are now quite common, and most contrast studies involve little risk. However, several types of studies involve added risks associated with the introduction of the contrast "dye" into the body.[6] The dangers of contrast-medium examinations include the potential for inducing a stroke or causing nerve damage when the dye is being introduced into the body. Some of the higher risk contrast x-ray examinations are listed in Table 4.

Since contrast examinations make up such a small percentage of the total number of x-ray examinations, the overall risks to the population from diagnostic x rays are still due primarily to radiation effects.

TABLE 1: AVERAGE EFFECTIVE DOSES OF COMMON X-RAY EXAMINATIONS (SOMATIC DOSES)

A. **High Dose Examinations** (more than 125 mrad or mrem per average examination)
Mammography (breast examination)
Upper GI Series (barium swallow)
Thoracic spine (middle or dorsal spine)
Lower GI Series (barium enema, colon examination)
Lumbosacral spine (lower spine)
Lumbar spine (lower back)

B. **Medium Dose Examinations** (25-125 mrad or mrem per average examination)
Intravenous Pyelogram (IVP, exam of kidney, ureter, and bladder)
Cervical spine (upper spine)
Cholecystography (gallbladder examination)
K.U.B. (kidney, ureter, or bladder examination)
Skull
Lumbo-pelvic (examination of pelvis and lower spine)

C. **Low Dose Examinations** (less than 25 mrad or mrem per average examination)
Chest
Hip or upper femur (hip or upper thigh examination)
Shoulder
Dental (whole mouth or bitewing examination)
Extremities (feet, hands, forearm, etc.)

See Appendices A, B, and C for more details.

TABLE 2: AVERAGE DOSE TO THE REPRODUCTIVE ORGANS FROM COMMON X-RAY EXAMINATIONS (GENETIC DOSES)

A. **High Dose Examinations** (exposure of male gonads, more than 200 mrad per average exam)

Lower GI (barium enema, colon exam)

Intravenous Pyelogram (IVP, exam of kidney and ureter)

Lumbar spine (lower back)

+Lumbo-pelvic (exam of pelvis and lower spine)

+Hip or upper femur (exam of hip or thigh)

B. **Medium Dose Examinations** (exposure of male gonads between 10-200 mrads per average exam)

Upper GI (barium swallow)

*Cholecystography (gallbladder exam)

Thoracic spine (middle spine)

*Upper GI (barium swallow)

Abdomen

K.U.B. (kidney, ureter, and bladder)

C. **Low Dose Examinations** (exposure of gonads less than 10 mrad or mrem per average exam)

Cervical spine (upper spine)

Skull

Shoulder

Chest

Dental (whole mouth or bitewing)

Extremities (hands, feet, forearms, etc.)

See Appendices A and C for more details.

+ High category for females

* Medium category for females

TABLE 3: HIGH DOSES TO THE UTERUS AND FETUS

The following examinations of a pregnant woman may expose her unborn child to more than 4000 mrad or mrem:

Abdominal aortography (examination of the main arteries in the abdominal region)

Lower GI Series (barium enema)

Celiac angiography (examination of the blood vessels in the abdominal cavity)

Upper GI Series (barium swallow)

Hysterosalpingography (examination of the uterus and oviducts)

Pelvimetry (examination to measure the size of the pelvis)

Placentography (examination of the placenta)

Renal arteriography (examination of the kidney)

Urethrocytography (examination of the kidneys, urinary tract, bladder, and urethra)

Cystogram (examination of the bladder)

See Appendices A and B for more details.

TABLE 4: HIGH RISK CONTRAST X-RAY EXAMINATIONS *

Type of Study	Method	Uses
Bronchogram	dye injected into lung bronchi (air passages)	outlines bronchial tree
Cerebral angiogram (arteriogram)	dye injected into carotid and/or vertebral arteries in neck	outlines blood vessels in neck and brain
Coronary angiogram (arteriogram)	dye injected into chambers of heart	outlines heart chambers, valves, and surrounding arteries and veins
Pneumoencephalo-gram (PEG)	air injected (as per myelogram); air rises into brain	outlines chambers and surface of brain
Pulmonary angiogram (arteriogram)	dye injected into pulmonary arteries as they leave heart	outlines blood vessels (arteries and veins) in lungs

* Source: Arthur Levin, *Talk Back To Your Doctor* (New York: Doubleday, 1975).

A Word of Caution!

It is assumed that each and every x-ray examination adds directly to an individual's risk in proportion to the "effective dose" to which he or she is subjected. Although it is desirable to keep the total radiation dose received by a person as low as possible, it is important to realize that the risks associated with each proposed x-ray examination must be considered in light of the potential benefits of that same examination. Thus if a given x-ray examination appears necessary, a patient's past history of radiation exposure should not influence the decision to conduct that examination.

Don't forget to consider the substantial benefits of diagnostic x rays. GI examinations as part of an annual routine executive physical may be completely unnecessary. On the other hand, in the presence of certain symptoms, the information obtained from a relatively "high" risk GI examination may save your life! In such cases the risk of *not* having a GI examination is greater than the risk of the examination itself.

Avoid routine examinations or prescribing x-ray examinations for yourself. However, do not avoid an x-ray examination if your physician can adequately explain why there is a real need for it!

In spite of some of the shortcomings of certain dentists, physicians, and radiologists discussed in this book, a qualified professional is probably a better judge of when x rays are needed than you are. You have a right to expect any professional with whom you are dealing to provide complete answers to your questions about the need for a proposed x-ray examination. If, after reading Chapter 14, you feel that the practitioner's reasoning and diagnostic techniques are sound, you should abide by his or her judgment. If you are not confident about this judgment, by all means seek an independent one from another professional.

CHAPTER 14
HOW TO MINIMIZE
YOUR EXPOSURE
TO X RAYS

To minimize your exposure to x rays and still obtain legitimate benefits from x-ray diagnosis, you should be concerned with three major areas. First, you must discuss the risk-benefit question openly with your physician or dentist. A proposed x-ray examination might not be necessary. Diagnostic x rays should be used only when the information gained will more than compensate for the potentially harmful effects of radiation. Second, you must assess whether the x-ray equipment and facilities are adequate. Finally, you must be in a position to evaluate some of the more visible techniques and procedures employed by the people who operate the x-ray equipment. The amount of exposure can be decreased significantly if proper techniques and adequate equipment are used.

The information in this chapter is intended to help you evaluate your physician, dentist, radiologist, and x-ray technologist and their facilities.

The Do's and Don'ts of Submitting to X rays

1. ASK THE PHYSICIAN, DENTIST, OR RADIOLOGIST WHO PROPOSES AN X-RAY EXAMINATION TO EXPLAIN WHAT IDENTIFIABLE BENEFIT WILL RESULT FROM IT.

The following examinations are of special concern because they involve relatively high overall radiation doses:

1. Mammography (breast examination)
2. Gastrointestinal examinations (upper or lower)
3. Thoracic Spine (middle or dorsal spine)
4. Lumbosacral spine (lower spine)
5. Lumbar Spine (lower back)
6. Intravenous Pyelogram, IVP (kidney, ureter, and bladder)
7. Cervical Spine (upper spine)
8. Cholecystography (gallbladder)
9. K.U.B. (kidney, ureter, or bladder)
10. Skull
11. Pelvis
12. Hip or Upper femur (hip or upper thigh)
13. Any fluoroscopic procedure (used to visualize motion)

The following examinations involve a significantly smaller dose on the average than those listed above:

1. Chest
2. Shoulder
3. Dental
4. Extremities (hand, foot, elbow, knee, etc.)

The need to question a practitioner about the benefit of a proposed examination in the low dose group is less important than in the high dose group. However, professional organizations like the American College of Radiology, the American Medical Association, and the Environmental Protection Agency,[1] have recommended that no x rays should be taken unless clear clinical reasons exist which indicate that they will contribute to a diagnosis.

In addition to radiation risks there is potential for harm associated with certain diagnostic examinations in which contrast dyes are used. The following "contrast" examinations involve significantly more risk than others:

1. Bronchogram (outlines bronchial tree)
2. Cerebral angiogram (outlines blood vessels in the neck and brain)

3. Coronary angiogram or arteriogram (outlines heart and surrounding arteries and veins)

4. Pneumoencephalogram (outlines chambers and surface of the brain)

5. Pulmonary angiogram (outlines blood vessels in the lungs)

You should make an extra effort to determine the medical need for any of these examinations before submitting to them.

---

2. ASK THE PRACTITIONER IF IT IS POSSIBLE TO USE THE RESULTS OF PREVIOUS X-RAY DIAGNOSES INSTEAD OF TAKING NEW EXPOSURES.

Sometimes a repeat examination is necessary to observe changes in a condition. However, physicians and dentists may fail to use current films because they do not trust previous studies or because they don't want to bother sending for them, or because they are not informed of their existence. Keeping a complete record of your own x-ray history on the form in Appendix D can help you keep your practitioner informed about available records.

Some hospitals and medical groups make a practice of not accepting x rays from other facilities. This practice should be questioned.

---

3. YOU SHOULD EXPRESS SPECIAL CONCERN TO YOUR PRACTITIONER ABOUT THE NEED TO X-RAY CHILDREN.

The potential for undesirable effects is greater in younger patients for several reasons. Children depend on normal cell division to support their rapid growth, and radiation is known to impair cell reproduction. Children have a longer remaining life expectancy, so that effects like leukemia and cancer have more time to manifest themselves in later years. Since children are small, it is often more difficult to keep their reproductive organs out of the primary x-ray beam. Finally, children are potential child-bearers, and genetic damage is of special concern for this group.

---

4. IF YOU ARE A WOMAN AND THERE IS ANY POSSIBILITY OF YOUR BEING PREGNANT, INFORM YOUR PHYSICIAN OR DENTIST. DON'T WAIT TO BE ASKED!

X rays which place a ripening egg or developing embryo or fetus in the main beam should not be taken unless absolutely necessary. X-ray examinations of a pregnant or potentially pregnant woman should be delayed if at all possible. An x-ray examination to diagnose early pregnancy should always be refused since there are other pregnancy tests available. Some practitioners will not ask you about a possible pregnancy for fear of embarrassing you, so it may be up to you to offer the information.

---

5. IF YOU ARE PREGNANT YOU SHOULD AVOID ALL X-RAY EXAMINA-TIONS OF THE LOWER BACK OR ABDOMINAL REGION UNLESS THERE ARE STRONG INDICATIONS OF A SERIOUS CONDITION.

A list of examinations which expose unborn children to the highest doses is included in Chapter 13, Table 3. In addition, some of the other examinations to be wary of if you are pregnant (because they may include the uterus and unborn child in the main beam) are:

1. Lower GI (barium enema)
2. Lumbar Spine
3. Thoracic Spine
4. Lumbo-pelvic
5. Cholecystography (gallbladder)
6. Intravenous Pyelogram, IVP (kidney and ureter)
7. Upper GI
8. Hip or upper thigh
9. Any other examination of the abdominal region.

You should avoid all x-ray examinations conducted during pregnancy to visualize the developing baby or size of the pelvis, if your physician performs them on a *routine* basis.

---

6. IF POSSIBLE, HAVE A PHYSICAL EXAMINATION AND A DENTAL CHECKUP BEFORE A PLANNED PREGNANCY.

Hormonal changes during pregnancy increase the chances of certain health problems occurring. For example, women are especially prone to developing gum and bone disease during pregnancy. Knowing whether or not you are in good physical condition ahead of time is worthwhile. Then if a physician or dentist recommends a diagnostic x ray on the basis of a physical examination, you can complete it before your pregnancy.

7. IF OTHER MEANS OF OBTAINING X RAYS ARE AVAILABLE, AVOID MOBILE UNITS.

Several knowledgeable organizations are now urging elimination of mass chest x-ray screening of the general population using x-ray vans. In spite of this fact, a few local TB associations are still promoting these programs. Modern diagnostic techniques such as the tuberculin skin test, and the decline in the incidence of tuberculosis and pulmonary lung disease, have made most mass chest x rays for population screening unnecessary. Besides, the mobile x-ray units used often produce miniature films, a process which usually involves significantly more exposure than conventional x-ray machines.[2] Don't decide on your own to have an x-ray examination. If a chest or any other x-ray examination seems necessary to you, consult with a physician first rather than prescribe it for yourself.

8. IF YOU ARE A WOMAN UNDER 50 YEARS OF AGE WITH NO SYMPTOMS OR FAMILY HISTORY OF BREAST CANCER, DO NOT SUBMIT TO MAMMOGRAPHIC SCREENING ON A ROUTINE BASIS.

As of August 1976, the National Cancer Institute (NCI) no longer recommends the use of mammograms as a screening tool for symptom-free women under 50 years of age with no family history of breast cancer. However, all women should practice breast self-examination on a monthly basis.

X-ray screening on an annual basis is recommended for women over the age of 35 who have a family history of breast cancer or one or more of the following symptoms: pain, lumps, discharge, or previous breast tumors. These women should consult with a physician.

Women over 50 with no symptoms or family history of breast cancer can receive free breast cancer screening at one of 27 breast cancer detection clinics. Women between the ages of 35 and 50 with no symptoms or family history of breast cancer may also receive free breast cancer screening which, if they so request, does not include a mammographic x-ray examination. (See Appendix F for a list of clinics and their locations.)

9. AVOID FLUOROSCOPY IF YOUR PHYSICIAN ACKNOWLEDGES THAT ORDINARY X-RAY FILMS WILL PROVIDE ADEQUATE INFORMATION.

The use of the fluoroscope is similar to taking an on-the-spot movie, and it exposes you to the x-ray beam for a relatively long time. The x ray is viewed directly on a screen, and may be recorded on film or videotape. Thus, a fluoroscope should only be used to visualize movements or change. Although newer fluoroscopes are equipped with image intensifying devices to give better images with less exposure to you, the exposure can still be considerably higher than that acquired from standard x-ray film. Standard x-ray films usually provide clearer images anyway, and have the advantage of providing a permanent record. However, when a fluoroscopic examination is being conducted, a physician will often expose the necessary radiographic films at the same time.

---

10. QUESTION THE NEED FOR ROUTINE PREEMPLOYMENT X-RAY EXAMINATIONS.

Employees who handle food or work with people are often required to have chest x rays to screen for tuberculosis. Ask if a tuberculin skin test or other tests can be substituted. If your skin test is positive, a chest x-ray examination may be necessary.

Employers for jobs requiring heavy lifting are in the habit of requiring lower back x rays which often involve placing the reproductive organs in the main x-ray beam. This procedure involves more radiation exposure than a chest x ray. One of the major purposes of lower back examinations is to protect the employer from a legal negligence suit should you develop lower back troubles on the job because of congenital deformities. If you must submit to lower back x rays for the record, previous x rays should be acceptable. If none are on file, go to a well-equipped facility supervised by a radiologist rather than to a mobile unit or physician's office. Men who are still of reproductive age should request a lead shield to protect their reproductive organs (testes).

---

11. REFUSE TO SUBMIT TO DENTAL X RAYS AS PART OF EVERY ROUTINE CHECKUP UNLESS YOU HAVE SPECIAL PROBLEMS OR UNLESS YOUR DENTIST CAN JUSTIFY THE NEED FOR THESE X RAYS.

The Environmental Protection Agency has stated that radiographic examinations should not be a standard part of every dental examination. Prudent dentists only require full mouth x rays (16 to 18 films with a conventional dental x-ray unit) every six to 10 years, and then expose single x-ray films in between when there are clear clinical indications of problems. (Each film consists of a 1¼ inch by 1⅝ inch paper-covered rectangle.) Additional exposures should be used for routine dental checkups only when there is rampant tooth decay in a child or some unusual condition that requires follow-up.

12. IF YOU CHANGE DENTISTS OR ARE REFERRED TO A SPECIALIST, RE-QUEST THAT YOUR NEW DENTIST OBTAIN ANY AVAILABLE DENTAL X-RAY RECORDS.

Previous records can help your new dentist understand your dental history and/or problems without adding to your x-ray exposure.

Checking Out the Equipment and Facilities

1. IN GENERAL YOU WILL RECEIVE SIGNIFICANTLY LESS RADIATION EXPOSURE AT A FACILITY WHICH IS UNDER THE SUPERVISION OF A FULL TIME RADIOLOGIST.

For example, the 1970 X-ray Exposure Study conducted by the Public Health Service indicates that examinations supervised by radiologists in private offices or hospitals typically involved less x-ray exposure than those delivered by non-radiologists in their private offices.[3] Mobile units with fluorographic equipment usually give about 5 to 10 times more exposure than the better types of facilities.

A specialist who uses his equipment and facilities full time is usually better able to afford up-to-date x-ray equipment and adequate facilities. There are exceptions to this rule, of course, so the best procedure is to check for some of the more obvious characteristics of a good facility outlined in the remainder of this chapter.

2. INQUIRE ABOUT WHETHER THE X-RAY FACILITIES HAVE BEEN IN-SPECTED BY ANY LICENSING AGENCIES OR PROFESSIONAL ORGANIZATIONS.

Radiation control agencies in most states conduct periodic surveys of x-ray facilities. Results of inspection surveys are sometimes open to the public, and you can find out how skin exposure rates from given x-ray tubes compare with those used at comparable facilities. Contact your State Radiological Health Agency for more information.

In addition, professional organizations like the American College of Radiology may review facilities and procedures as a way to help promote professional standards among their members. You can ask what organizations have reviewed the facilities and practices at your x-ray establishment.

---

3. AT A MEDICAL X-RAY FACILITY LOOK FOR OR ASK ABOUT THE PRESENCE OF A BEAM-LOCALIZING LIGHT AND AN ADJUSTABLE RECTANGU-LAR BEAM RESTRICTOR LOCATED IN FRONT OF THE X-RAY TUBE.

If the area of the x-ray beam is not restricted to the size of the film or smaller, you will be receiving needless exposure. For example, studies have indicated that proper restriction of the beam size can result in an average reduction of the significant dose to the reproductive organs by about 65 percent.[4]

The beam restrictor is called a collimator and works like the iris diaphragm in a camera, except the shutters which open and close are made of lead, and it produces a rectangular rather than a circular beam. The beam-localizing light is an ordinary light located behind the shutters. When the light is turned on, it is projected through the shutters and outlines the size and location that the x-ray beam will have.

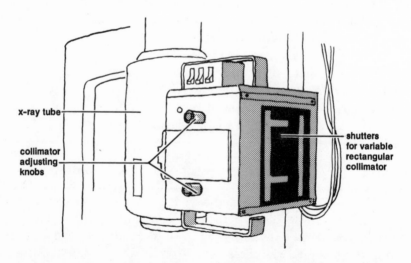

x-ray tube

collimator adjusting knobs

shutters for variable rectangular collimator

X-ray tube with shutters (collimators) for restricting the size of an x-ray beam

Thus, the operator can adjust the shutters and aim the x-ray tube properly before turning on the x-ray beam. *The shutters should be adjusted so the light beam is no larger than the film.* In many cases, particularly when x-raying children, it is desirable to limit the x-ray beam to an area smaller than the film. If it is obvious that the light beam is much larger than the film or area being examined, *point this out to the technologist before the x-ray beam is turned on.*

In compliance with the Radiation for Health and Safety Act of 1968, H.E.W. regulations require that machines presently being manufactured be designed so that the x-ray technologist cannot make an exposure unless the x-ray beam is collimated to, at most, the size of the film. Although the beam size is adjusted automatically in the latest machines, the light beam localizer is still used to outline the beam and position the x-ray tube and patient properly. Very old x-ray machines do not have adjustable collimators or a localizing light, but are usually equipped with several non-adjustable cones which fit over the front of the x-ray tube. These machines can be used safely if the machine operator is conscientious about using a cone which restricts the area of the circular beam to about that of the film and if the beam is aligned carefully. *Operators have been known to work without adequate collimation to avoid bad exposures due to misalignment. Do not allow yourself to be x-rayed under this circumstance.*

---

4. AVOID EXPOSURE TO OLD-FASHIONED FLUOROSCOPES.

Ask about the presence of image-intensification equipment before submitting to a fluoroscopic type of x-ray examination. Modern fluoroscopic equipment amplifies the x-ray image so that significantly lower exposures result. With older equipment the unamplified image is quite faint. The examiner must work in a darkened room and wear red goggles when the lights are on to adapt his or her eyes to the dark.

---

5. IF YOU ARE GOING TO RECEIVE AN INITIAL MAMMOGRAPHIC EXAMINATION, GO TO A FACILITY WHICH USES LOW DOSE MAMMOGRAPHIC EQUIPMENT.

A breast examination should require no more than one rad of skin dose per film when proper equipment and techniques are used. Xeroradiography, which works on the same principle as Xerox copiers do, is rapidly replacing equipment requiring conventional x-ray film. (See the Glossary for a brief description of this process.) It is a good low dose method. The equipment is easy to identify, because the technician is required to put an exposed cartridge into a large specially designed Xerox machine to obtain an image. No photographic films are developed.

If no xeroradiographic machine is available in your community, ask about the skin dose to the breast associated with the machine proposed for your examination. (High dose machines are often used with no intensifying screen surrounding the film and with slower speed industrial type x-ray film.) If the dose is more than one rad per exposure, try to find another facility with low dose equipment.

Some radiologists feel that certain higher dose techniques yield more diagnostic information than the low dose methods described here. If your radiologist feels this way, ask that it be used only if an initial low dose examination indicates that a more detailed, higher dose study is needed.

---

6. IF YOU ARE TO BE X-RAYED WITH A CONVENTIONAL DENTAL X-RAY UNIT, ASK FOR ONE WITH A LONG OPEN-ENDED, LEAD-LINED CYLINDER ON THE END RATHER THAN ONE WITH A SHORT, POINTED, PLASTIC CONE.

The newer type of lead-lined cylinder, which is about eight inches long, is designed to produce a narrow beam and a minimum of scattered radiation. Scattered x rays give you needless exposure without contributing to the quality of the image recorded on film. The exposure of the reproductive organs due to scattered radiation is about twice as much with the pointed plastic cone.

Your dentist can install long cylinders on his x-ray units without much additional expense. If he still uses pointed plastic cones, ask him to switch.

---

7. THERE ARE SEVERAL OTHER IMPORTANT DEVICES WHICH HELP MINIMIZE UNNECESSARY EXPOSURE WHICH ARE NOT LIKELY TO BE VISIBLE TO YOU. YOU MAY WANT TO ASK ABOUT SOME OF THESE THINGS IF THE OPERATOR HAS TIME.

A. *Fast film:* The use of ultraspeed x-ray film for medical and dental x rays reduces your exposure by two-thirds or more compared to slower films. In some cases slower films which give greater resolution must still be used to obtain adequate information for a diagnosis.

B. *Electronic timer:* The newer electronic timers take the guess work out of setting short exposure times, and thus minimize the number of retakes due to poor exposure.

C. *Metal filter:* The use of a metal filter, usually aluminum, is important in screening out low energy x rays which are absorbed by a patient but do not contribute to the quality of the x-ray picture. In a few special cases the filter is not used.

D. *Metal grid:* If a moving metal grid is set directly in front of the film when medical x rays are used, some of the scattered radiation which detracts from the image of interest is absorbed. This apparatus is called a Potter-Bucky diaphragm or "bucky" for short. (A metal grid is not desirable in all types of examinations.)

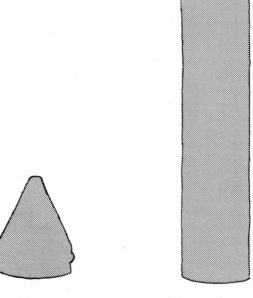

short, pointed dental cone   **long, open-ended, lead-lined dental cylinder**

E. *Image-intensifying screen:* 1. For conventional x rays other than those of the extremities (hands, feet, etc.) it is often advantageous to place intensifying screens directly around the film in the film holder while the film is being exposed to x rays. The screens absorb x-ray energy which passes through the film and gives off additional light which re-exposes the film. Since the image on the film contributed by the intensifying screens is not as sharp as the image caused by the main x-ray beam, it is not always desirable to use screens.

Checking Out the X-ray Machine Operator

1. IN TAKING MEDICAL EXPOSURES THE OPERATOR SHOULD MEASURE THE THICKNESS OF THE PART OF YOUR BODY WHICH IS TO BE EXPOSED AND CONSULT A TECHNIQUE CHART TO SET THE TUBE CURRENT, VOLTAGE, AND EXPOSURE TIME FOR EACH TYPE OF X RAY.

Operators who guess or hurry through this procedure are more likely to take poor exposures. Repeated exposures due to sloppy techniques cause needless exposure to radiation. If the patient has an unusual muscle or bone density or moves during an exposure, the exposure will have to be repeated anyway. Careful operators have a repeat rate of less than five percent.

2. A GOOD OPERATOR ALIGNS AND RESTRICTS THE BEAM CAREFULLY, ESPECIALLY WHEN THE GONADS MIGHT BE IN THE PRIMARY BEAM.

For example, by taking an x ray of the forearm with the patient sitting and the beam directed downward, the gonads can be included inadvertently in the direct beam. Proper alignment and use of a minimum beam size sometimes takes ingenuity. It is more difficult to keep the female ovaries than the male testes out of the primary beam. The operator cannot always avoid including a patient's reproductive organs in the direct x-ray beam. For example, hip, lumbar spine, and lower gastrointestinal tract examinations can all expose the gonads directly. *In many cases shielding the reproductive organs is possible, especially for men, and should be requested by all potential childbearers.* In some cases where the shielding obscures the organs of interest, shielding may not be possible. (See item 5.)

3. A GOOD OPERATOR USES EXTRA CARE IN ALIGNMENT AND A
SMALLER EXPOSED AREA OF FILM WITH CHILDREN.

Children's organs are closer together and they are more sensitive
to radiation damage. When exposing small children the operator should
either use a smaller film size and narrow the beam, or just expose a
small part of a larger film.

4. COOPERATE WITH THE OPERATOR DURING X-RAY EXPOSURES—DON'T
BREATHE OR MOVE A MUSCLE.

This cooperation helps prevent blurred images and the need for
"retakes." Since it is difficult for infants and young children to hold
still, a parent or another adult may be asked to hold a child to prevent a
blurred exposure. *If you are asked to hold a youngster, request a lead
apron. Refuse to help if you are pregnant.*

5. IF YOU ARE IN THE REPRODUCTIVE YEARS OR YOUR CHILD IS TO BE
X-RAYED, ASK FOR A LEAD SHIELD FOR THE REPRODUCTIVE ORGANS UNLESS
THE PRESENCE OF A SHIELD WILL INTERFERE SIGNIFICANTLY WITH THE
INFORMATION TO BE OBTAINED BY THE X RAY.[5]

Some examinations which directly expose the testes or male reproductive organs are listed below. Males of reproductive age or younger
should ask for a gonadal shield for the following examinations: [6]

Lower GI (barium enema)
Upper GI (barium swallow)
Intravenous Pyelogram (IVP, kidney, and ureter)
Lumbo-pelvic
Hip or femur
Lumbar spine
Abdomen
Sacrum and coccyx
Pelvis
Myelogram
Gallbladder

Even when the reproductive organs are not in the main beam,
shielding can help reduce the amount of scattered radiation absorbed

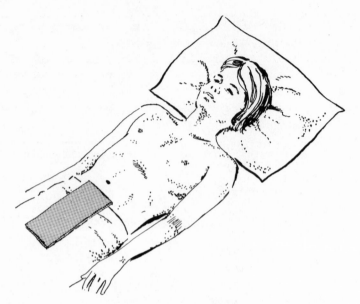

Lead panel-type contact shield

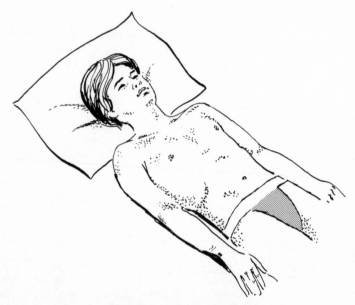

Scrotal cup in lead underpants

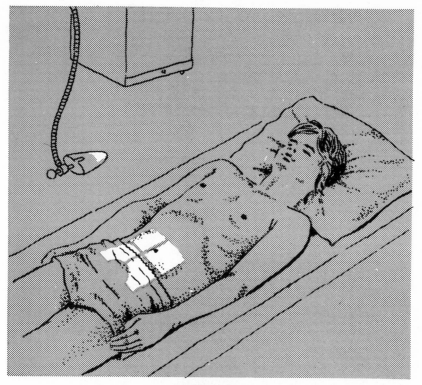

Shadow shield

by the reproductive organs. Shielding of the male testes is possible in almost all cases, although many times it is impossible to shield the female ovaries from direct radiation. (The examinations involving relatively high doses for the female ovaries are listed earlier in this chapter and in Table 3 of Chapter 13.) Shields are often so easy to use that the extra measure of protection is very worthwhile. Some practitioners and operators do not offer shielding to patients, to avoid frightening or embarrassing them. There are several types of shields including lead aprons, lead-lined panels, scrotal cups, flexible lead-lined drape cloths, and shadow shields.

Ask about them before being x-rayed. Remember, however, that shielding of the reproductive organs should be used in addition to good collimation and not as a substitute for it. For example, in a lower back examination of a male, an inch or two of extra collimation at the lower end of the film can make a tremendous difference in his testicular dose.

---

6. REQUEST THAT A LEAD APRON BE PLACED OVER YOUR CHEST AND LAP FOR DENTAL X RAYS IF YOU ARE A POTENTIAL PARENT AND YOUR REPRODUCTIVE ORGANS MIGHT BE IN THE DIRECT X-RAY BEAM.

In most dental examinations the reproductive organs do not lie in the path of the primary x-ray beam and a lead apron is not needed. However, a dentist who aims the x-ray beam down through the top front teeth may be including the reproductive organs in the main beam. In that case, you should request a lead apron for yourself or for your children (who are also potential parents).

Check List of Questions to Ask

A. Questions everybody should ask medical and dental personnel:

- What benefit should I expect from the proposed x-ray examination?
- Are there clinical indications that an x-ray examination is needed or is this a routine examination?
- Would you like to know about my previous x-ray examinations and would it be possible to use the results of any of them?
- Is this x-ray facility inspected by any licensing agencies?

B. Questions everyone should ask medical personnel:

- Why is fluoroscopy necessary in my case?
- Is this facility supervised by a full time radiologist?
- Are you able to adjust the size of the x-ray beam to that of the smallest appropriate film size?

Dental personnel:

- Do you use a long, open-ended lead cylinder?

C. Questions for young adults and children who are potential parents to ask medical and dental personnel:

- Will my reproductive organs be in the main beam, and if so, can you provide me with a lead shield for them?

D. Questions for women who are or may be pregnant to ask medical and dental personnel:

- Do you know that I may be pregnant?
- Can this x-ray examination wait until later in my pregnancy or just after my next menstrual period?
- Will my uterus be in the main beam, and if so, can you provide me with lead shielding for it?

CHAPTER 15
ANSWERS TO COMMON
QUESTIONS ABOUT
MEDICAL AND DENTAL
X RAYS

A friend once said in jest that the title of this book ought to be "What You Always Wanted to Know About X rays, But Were Afraid to Ask." Often I have found that people who write to me for advice about x rays are either shy about asking their questions, or they have not been getting satisfactory answers. Many physicians, dentists, and x-ray personnel do not know the answers to their patients' questions or are afraid to answer them.

Although the answers to most of the questions which I have been asked are contained in this book, so many people asked me the same questions that I wanted to share them with you.

1. I have heard that dental x rays are no more dangerous than spending an hour or so in the sun. Is that correct?

No, it is not correct because dental x rays, unlike the ultraviolet radiation from the sun, are capable of penetrating the body, so that they expose internal tissues. However, the skin exposure from dental x rays and the ultraviolet radiation received in a sunbath are comparable. This similarity in skin exposure has been a source of confusion for medical and dental professionals as well as for patients. Both sunlight and dental x rays serve to increase an exposed individual's chances of getting skin cancer, though skin cancer is rarely fatal.

Although dental x rays are more penetrating than sunlight, the only internal organs likely to be significantly exposed to dental x rays, which are prone to x-ray damage from this kind of exposure, are the thyroid gland located in the neck and the lens of the eye. If the effects of x rays are strictly proportional to the doses, dental patients may have some increased chance of developing cataracts or thyroid cancer sometime in the future. It appears from various studies that x rays do not induce cataracts at low doses, and thus, dental x rays are not likely to produce cataracts. Also, there is no direct evidence that dental x rays cause an increased risk of thyroid cancer or leukemia, but it is believed that any amount of radiation dose may cause increased chances of cancer or leukemia in proportion to that dose. The chance of developing cancer or leukemia from properly conducted dental x rays is very remote and thyroid cancer is rarely fatal. Therefore, even though the risks associated with dental x rays are greater than those associated with a sunbath, these risks are small compared to many everyday hazards which people take for granted, like smoking, riding in a car, or crossing streets.

2. I have just had a new baby and have been subjected to a chest x ray in the hospital. Will my milk be contaminated?

No, your milk cannot be contaminated by x rays. When a diagnostic x-ray beam passes through milk it may change the structure of a very, very small number of the milk molecules. These few molecules are not radioactive or considered harmful to young babies in any way. Most of the x rays in a beam will pass through the milk.

The notion that x rays might contaminate milk probably comes from the concern about the potential of radioactive fallout to contaminate mothers' milk and the bones of young children. In the late 1950s and early 1960s the United States and other countries were conducting atmospheric testing of atomic and nuclear weapons. Radioactive materials capable of emitting particles such as alpha, beta, gamma, or x rays were being released into the atmosphere. These materials would contaminate human food and thus find their way into the mothers' milk. One of these materials, known as strontium 90, was considered dangerous to humans because strontium is chemically similar to the calcium in milk and in bones. When a baby consumed contaminated milk, it was possible for strontium atoms to be incorporated instead of calcium atoms in its teeth or bones. Eventually each radioactive strontium 90 atom will eject a beta particle which can cause damage to the cells of a young child very similar to the damage caused by x rays. Fortunately, there is virtually no atmospheric testing being conducted any longer, and milk contamination is not now a significant threat to human health.

The use of nuclear scanning for diagnostic purposes involves the introduction of certain radioactive materials into the body, and some types of nuclear scanning examinations may cause contamination of the milk of breast-feeding mothers. Nursing mothers should find out more about this possibility before undergoing any nuclear scanning procedures.

3. I have had lots of x-ray examinations in the last few years. Can you tell me if I have had too many? Should I refuse to have any more?

It is not possible to say that you have or have not had too many x-ray examinations for your health. There is no firm evidence of a threshold effect for diagnostic x rays; it is believed that the danger associated with each x-ray examination is in direct proportion to the dose received from it. Thus, the overall risk to your health from diagnostic x rays is considered to be roughly proportional to the sum of "effective doses" of radiation your body has received for all your x-ray examinations. The *additional* risk contributed by each subsequent examination is probably the same, regardless of the number of previous x-ray examinations you have had.

If there is a reasonable expectation that a proposed diagnostic x ray will yield valuable diagnostic information which will influence the course of your treatment, it makes no sense to refuse it—even if you have had many x rays in the past. On the other hand, an x-ray examination which isn't really necessary is not wise, even if you've never had a previous one. Sometimes a proposed examination is a repeat of a similar x-ray study conducted recently. You should refuse such repeated examinations unless your prior films are not available or unless the physician can convince you that there is a real need for a new study.

4. Should my dentist place a lead apron on my lap when he x-rays my teeth?

Only in special circumstances. The lead apron is only valuable in cases where the reproductive organs of a potential parent or the uterus of a pregnant woman might be in the main x-ray beam. Most dental examinations do not place these organs in the main beam, although it is possible to do so during an x-ray examination of the top front teeth. In most dental procedures the small x-ray exposure which the reproductive organs receive does not come from the direct x-ray beam. These x rays are scattered from objects in the room or other parts of the body.

Many dental patients who are concerned about their x-ray exposure view the use of a lead apron as tangible proof that their dentist is appropriately cautious about x-ray use. Although the lead apron is psychologically reassuring, it does not actually reduce the dose to the gonads for those dental examinations which do not place the reproductive organs in the main beam. Even though the lead apron protects the reproductive organs from direct radiation, it can scatter indirect radiation into the reproductive organs causing slightly more exposure of the gonads or reproductive organs. Actually the scattered radiation exposure to the reproductive organs is usually so small that the presence or absence of a lead apron makes essentially no difference to the patient.

5. My daughter just had x rays of her abdomen. Shouldn't her reproductive organs have been shielded? Will she be sterile?

Unfortunately, it is not usually possible to shield the ovaries of girls or women without obscuring what the physician is looking for on the x-ray film. Shielding is only important in cases where the ovaries might be in the primary or direct x-ray beam, e.g., examinations of the

186

abdomen, lower back, or hip. In other cases good collimation or beam restriction is the best protection against unnecessary radiation to the gonads.

Since it is difficult to shield the ovaries of a young girl, it is especially important to establish need for an x-ray study of the abdominal area or lower back. When these types of x rays are needed, parents should be certain that proper collimation is being used (See Chapter 14 for details).

An x-ray examination which places a young girl's ovaries in the main beam may slightly increase her chances of passing on defective genes to future generations. However, it is important to understand that this individual risk is very small. The real long-term danger of genetic damage from radiation occurs when large segments of the population are exposed to this small risk. For then the number of affected individuals in future generations can become significant. Although large doses of radiation can cause sterility (the inability to produce children), it is absolutely untrue that the low doses of radiation associated with any diagnostic procedure can cause sterility.

6. I am pregnant with my first child. My doctor ordered an x ray of my pelvis to determine its size. Now I have read that x rays during pregnancy can cause deformities and leukemia. Will my baby be deformed or have leukemia?

It is not likely that your baby will be deformed or have leukemia. The effect of a pelvimetric x ray late in pregnancy is to cause a slight increase in the chance that your baby will have leukemia during childhood. *This is very different from stating that your child will have leukemia.* Pelvimetric x rays usually occur in the later stages of pregnancy, and x rays late in pregnancy, after the fetus is fully formed, do not result in deformities.

However, serious questions have been raised about the value of routine pelvimetric examinations (See Chapter 6). Since the fetus is considerably more sensitive to radiation effects than an adult is, it is vital for expectant mothers to ascertain the real need for any x-ray examination of the lower spine or abdomen.

7. My doctor ordered an x ray of my abdomen last month. I just found out I was pregnant and I am afraid my child has been damaged. Should I consider having an abortion?

Probably not. Abdominal x rays received during the early stages of pregnancy may increase the probability that a developing embryo will undergo a spontaneous abortion (miscarriage) or will develop a deformity. However, most experts agree that unless an expectant mother receives an unusually large number of high-dose abdominal x-ray examinations (See Chapter 14) the risks to the child are not large enough to warrant a therapeutic abortion. The chances of damage to your child are reasonably small in most cases, and even if your child is born with some type of abnormality, it would be impossible to trace it definitely to your x-ray exposure. This is because there are many other factors which could also contribute to an abnormality.

8. I injured my back a year ago and have been to several doctors and specialists. Each time I see a new doctor he wants a new set of back x rays. Are all these repeats necessary?

Many physicians take their own x rays even when results of identical x-ray studies are already available. Sometimes repeated studies are needed to monitor changes in a condition or because previous films are of inferior quality, but it does appear that some physicians order new x-ray studies just because it is easier than making the effort to acquire previous results. Some patients have attempted to obtain their previous films to carry with them when visiting a new specialist. However, many doctors and hospitals are reluctant to release their films to patients.

In legal disputes, district courts have usually ruled that x rays are the property of the prescribing physician or the x-ray facility, although I know of at least one case in which x rays were judged to belong to the patient. However, patients can attempt to keep a complete record of their x-ray studies, including information about the type, purpose, location, and date of each examination. In this way they can inform their physicians and dentists about the existence of any prior studies which might be of interest (See Appendix D).

9. I have been suffering from a chronic illness. Every time I am admitted to the hospital for tests they insist on taking another chest x ray. Why is this done?

Many hospitals have a rule that routine chest x rays must be taken before a patient can be admitted to the hospital. Hospitals claim that these x rays are necessary primarily to detect unsuspected cases of tuberculosis and other communicable respiratory diseases which might spread to other patients. There does not seem to be any good evidence that the incidence of these communicable diseases justifies the widespread use of routine chest x rays. For example, except among people subjected to over-crowded or unsanitary living conditions, the incidence of tuberculosis in the United States is currently very low. The rigidity with which this hospital rule is applied can be irritating and expensive to patients who undergo several chest x rays taken at the same hospital during a short period of time. Hospital personnel seem especially reluctant to even check the results of their previous chest examinations.

A chest x ray is legitimate for a patient about to receive surgery to ascertain whether or not he or she can safely receive a general anesthetic. However, there is no good reason for x-raying every patient who is admitted to the hospital.

10. My doctor ordered an extensive series of x-ray examinations. Since nothing unusual was found on them, I suspect that the tests were unnecessary. What do you think?

When an extensive *and* expensive series of x-ray examinations yields no diagnosis, patients often feel frustrated and cheated. Sometimes a patient's symptoms lead a physician to suspect that they might be the result of a condition which an x-ray study can reveal. If the x-ray study is negative the physician is able to "rule out" a specific disease and look for other causes of illness in the patient. Thus negative results on an x-ray examination do not always mean the examination was unnecessary. However, ruling out can be carried to unproductive extremes. Some physicians order extensive and apparently unrelated x-ray examinations on the basis of a minor symptom or complaint. One technician referred to these extensive sets of procedures as "fishing trips." In many of these cases it would be more prudent, less expensive, and less risky (but more time-consuming) to rule out one condition before investigating another. Thus a patient could be sent for one x-ray examination at a time, and the physician would be able to consider the results of each examination before ordering another one.

11. Are there any drugs or vitamins I can take to counteract the effects of x rays?

There are no known substances which can neutralize the effects of x rays or other ionizing radiation after an individual has been exposed. However, some research is being done on chemicals which may minimize the damage to cells if introduced into the body before radiation exposure. To understand how these chemicals probably act to minimize the effects of radiation, it is helpful to know more about how radiation damages cells.

It is currently believed that when one of the DNA molecules contained in the nucleus of every human cell is damaged by radiation, the potential increases for the cell to become a source of cancerous growth some years after the damage occurs. Typical cells contain a relatively large proportion of water (H_2O) and other molecules. DNA molecules which control cell reproduction make up a minute proportion of all of the molecules in the cell. Thus, there is very little chance that an x ray passing through a cell will hit and ionize a DNA molecule. However, when radiation passes through a cell many H_2O (water) molecules may be split into H^+ and OH^- ions or into free radicals. Both the OH radical and ion react very easily with other substances in cells including DNA molecules. It is believed that the main source of damage to DNA molecules results from the reaction of the OH radicals and ions with them. If the cell is bathed in any type of chemical which serves as an antioxidant and reacts easily with the OH radicals or ions, it is possible to neutralize these OHs before they do damage to the cellular DNA. A number of substances, including vitamin E, act as antioxidants in human cells.

Future research could show that if patients receive enough of an antioxidant such as vitamin E *before* a proposed high dose x-ray study, the harmful effects of radiation might be minimized, although at present there is no firm evidence for this effect.

12. My doctor claims x rays are harmless. I find this hard to believe when he so carefully protects himself with lead or leaves the room even though I'm the one getting direct exposure. What do you think?

It is recommended procedure for a physician, dentist, or technician to leave the room or wear a lead apron while conducting an x-ray examination. This is reasonable because people who operate x-ray machines often work with x rays on a full-time basis. They receive small radiation doses during each examination because x rays scattered from objects in the room can be absorbed by a machine operator. The dose from each examination must be added to the doses from the hundreds of examinations a machine operator conducts each year. The habit which machine operators have of protecting themselves is evidence that they understand that x rays are not completely harmless, although the potential for harm due to the scattered x rays from a single examination can be quite small.

13. My grandson died of leukemia recently. His dentist had been x-raying his teeth every six months. I am sure the excessive x rays caused the leukemia. Do you agree?

It is possible but very unlikely that routine dental examinations of a young child would result in leukemia. The amount of red bone marrow exposed during a dental x-ray procedure is extremely small, so that the chances that your grandson's leukemia was caused by the dental x rays are probably very low indeed.

It is important to realize that a single case of leukemia cannot be traced back with certainty to radiation exposure. This is because leukemia can occur in people who have had no x-ray exposure. All we can say is that x-ray exposure increases a person's statistical odds of developing certain diseases.

Even though the dental examinations may have increased your grandson's leukemia risk only very slightly, the major hazard of dental x rays is probably economic. Many dentists who conduct regular x-ray examinations do not check the condition of the teeth first, and this can cause needless expense (See Chapter 5).

14. I have a cancerous tumor and my doctor wants me to have x-ray therapy. I am deathly afraid of receiving too much radiation. What should I do?

If you feel that your doctor's diagnosis is sound, then I suggest you follow his or her advice. If your cancerous tumor has the potential of developing, it may become a threat to your life. In this case the immediate benefits of destroying the tumor with radiation probably outweigh the risk that you might develop another type of cancer or leukemia from the radiation exposure sometime in the future. Unfortunately the doses used in therapy must be much larger than those used in modern diagnostic x-ray procedures. There are some unpleasant side effects of radiation treatments such as nausea and loss of hair. These should be discussed with your physician, but they should not deter you from submitting to radiation therapy if it has the possibility of saving your life.

15. In 1975 I entered the hospital for a hemorrhoid operation and had the routine chest x ray, but because it didn't take properly, I had two exams in a row—front and left side views. Three months later I again had surgery, this time a cancerous lump on the left side of my left breast. I have no way, of course, of knowing if this had any connection, but it keeps bothering me that perhaps it did cause me the loss of a breast.

It is not considered possible for any type of x-ray procedure to cause breast cancer three months after the exposure. In fact, breast cancer has a relatively long latent period, and studies have shown that even groups of women who have received large doses of radiation do not start getting radiation-induced breast cancers until 15 to 20 years after their exposure to radiation.

16. I have read that regular x-ray mammograms (x-ray studies of the breast) in younger women may induce more breast cancers than they detect. I have never had a mammogram but I have had a number of

chest x rays which have exposed my breasts to x rays. Am I more likely to get breast cancer?

The exposure received by the breast from an ordinary chest x-ray examination is extremely small. Thus it is very unlikely that chest x rays have any measurable effect on breast cancer. On the other hand, the radiation dose to the breast from a mammographic examination may be as much as 1000 times greater. Thus the risk of breast cancer from routine mammography is of more serious concern. Don't confuse these two very different kinds of x-ray examinations.

If you are a woman of any age who needs a chest x-ray for a good reason, don't let the concern about radiation-induced breast cancer deter you from having one.

APPENDIX A: UNITS OF RADIATION EXPOSURE AND DOSE

In order to understand how x rays interact with people, it is necessary to have some understanding of the ways in which scientists measure the quantity of x rays to which a person is exposed as well as the amount of x-ray energy or dose that person absorbs. Once the quantities associated with exposure and dose are defined, it is possible to discuss the potential effects of certain exposures or doses on human health.

Defining appropriate units to describe the effects of radiation has proven difficult for scientists ever since the discovery of ionizing radiation. As a consequence, three related units of radiation, each slightly different from the others, have been defined. Respectively known as the roentgen, rad, and rem, they are often used interchangeably in describing x rays that are striking an individual. This appendix is an attempt to describe important distinctions between these units of radiation.

Exposure and Dose—What Are They?

The effects of diagnostic x rays on the human body depend in a complex way on a number of factors such as the distribution of energies of x-ray photons in the beam, the total intensity or quantity of radiation, the distance between the x-ray tube and the individual being x-rayed, the type and location of tissues and organs in the main beam, as well as the age and sex of the person being examined.

In trying to understand the relationship between the properties of a given amount of radiation and the biological effect on an individual, the concept of exposure was developed. Exposure is a measure of the number of electrons which are torn away from molecules when a beam of radiation passes through air. A positively charged air molecule and the negatively charged electron removed from it are referred to as an *ion pair*. The concept of radiation exposure is very convenient since it is easy to place an ionization chamber containing air in the path of radiation and record the number of ion pairs produced by the beam electronically. The common unit of exposure is the roentgen, named after Wilhelm Roentgen who discovered x rays in 1895. One roentgen is the amount of radiation necessary to produce 1600 trillion ion (electron-molecule) pairs in one kilogram of air (1 kilogram=2.2 pounds).

Although the exposure in roentgens is an easy quantity to measure, it is not always a good indicator of the effects of radiation on an individual. This is because tissue, bone, and other materials contain different types of molecules and have different densities from air. Furthermore, the processes by which ionizing radiation like x or gamma rays, beta particles, or alpha particles lose their energy in passing through matter is different for each type of radiation. The energy absorption process also depends on the quality or energy distribution of the radiation as well as on the type of exposed tissue or bone.

The concept of *absorbed dose* was developed as a measure of the amount of energy dumped by incident radiation into a gram of material. The dose absorbed by a gram of skin or muscle can be much less than that of a gram of bone placed in the same x-ray beam. This is because the heavy atoms of calcium in bone absorb x rays more easily than lighter elements abundant in tissue. Thus x rays pass through tissue more easily and don't leave as much energy behind.

Absorbed dose, although harder to measure, is probably a better indicator of the biological impact of radiation than exposure. The most common unit of absorbed dose is the rad. Because the outer layers of material absorb x rays readily the exposure inside a person's body will be less than the exposure at the skin. The absorbed dose in the outer layers of skin is often referred to as the *skin dose* while x-ray energy deposited in a gram of bone, tissues, or an organ at a certain location inside the body is referred to as the *depth dose* at that location. High energy x-ray photons have much more penetration power, and the depth dose corresponding to a relatively high energy (100 kVp) diagnostic x-ray beam can be hundreds of times greater than that of a low energy diagnostic x-ray beam (20 kVp).

Since the exposure in roentgens is the same as the absorbed dose in rads at the surface of soft tissue for medical x rays, these units are sometimes used interchangeably. *However, in most situations a measurement of x-ray exposure in roentgens made with an ion chamber at the surface of the body is not the same as the dose or energy absorbed at various points in the body in rads.* When other types of radiation like alpha particles or neutrons are present, a unit called the rem is used. Rem stands for Roentgen Equivalent Man and describes the potential for biological damage which results from a given dose in rads of radiation other than x rays. *However, for x rays, rad and rem always have the same value and can be used interchangeably.*

At present most researchers studying the biological effects of ionizing radiation on humans assume that its effects on a particular location in a body can be directly related to the absorbed dose in each gram of material at the location of interest. However, some effects of interest may be related to the time period in which the dose is observed, or *dose rate,* as well as the total absorbed dose. Very high dose rates are known to do more somatic damage to individuals than the same total dose delivered more slowly. The relationship between genetic and somatic damage and dose rates for lower doses is not well established, and it is safer in setting radiation standards to assume that biological effects depend only on total absorbed dose (except at very high dose rates).

The concepts of absorbed dose and dose rates are very powerful and useful. Our present state of knowledge is such that if an individual absorbs a known dose at each location in his body within a specified amount of time, it is possible to set upper limits using the linear hypothesis on the effects the radiation will probably have on his health.

APPENDIX B: EXPOSURES TO EMBRYO AND FETUS

Range of Possible Exposures of an Embryo or Fetus in Various X-ray Diagnostic Procedures of the Pregnant Woman *

Diagnostic Radiography	Dose Range in Millirads
Abdomen screening	100-2000 depending upon whethcr there are multiple films and/or fluoroscopy
Abdominal aortography *	6,000-20,000
Amniography	100 or more depending upon techniques used
Barium enema *	350-6,000 (higher with fluoroscopy and spot film)
Carcinoma of the cervix	To 6,000 if fluoroscopy is used
Cardiac series	5-50, higher with fluoroscopy
Celiac angiography *	2,000-20,000 (with fluoroscopy)
Cephalopelvimetry	See pelvimetry
Chest	Single view 1-10 or less; 5-70 or to 2,000 with fluoroscopy
Cholangiography	20-200 (films) or to 2,000 (fluoroscopy)

* Denotes procedures which are high risk to the fetus because they may deliver 4 rads or more to it.

Cholecystography	20-200 (films) or to 2,000 (fluoroscopy)
Cystogram (excluding urethra)	500-1,000
Colon	See GI series, lower tract
Dacryocystography	1-10
Dental series	0 with lead and rubber apron
Esophagram (esophagogram)	To 500 with fluoroscopy
Extremities	Negligible (1 mrad)
Femoral arteriography	1,000+
Fetal age	500-1,000
Fetometry	100-300
Gallbladder	See cholecystography
GI series, lower tract *	350-6,000, higher exposure with fluoroscopy
GI series, upper tract	100-2,000, higher exposure with fluoroscopy
Hip (spine and buttocks)	300-700
Hysterosalpingography *	1,200-6,000. Rarely involves pregnancy, upper exposure levels with radiopaque and fluoroscopy
Ileo-cecal study	100-2,000, higher exposure with fluoroscopy
Intestine, small study	100-1,000 with multiple films and fluoroscopy
K.U.B. (kidney, ureter, and bladder)	200
Long bone series	50-200
Mastoid areas	25-100
Myelography	Minimum of 2,000 spot films and fluoroscopy
Nephrotomography	500-2,000

* Denotes procedures which are high risk to the fetus because they may deliver 4 rads or more to it.

Obstetric examinations	500-2,000 if laterals and stereo included
Ocular foreign body detection	1-10
Pelvic pneumography	300
Pelvic, lower abdomen screen	80-500
Pelvimetry *	600-4,000 (usually toward term)
Petrous pyramids	1-10
Placenta praevia	500-1,000 with contrast injection
Placentography *	300-7,500 with contrast injection
Pneumoencephalography	2-20
Polytomography (both ears)	3-30
Prenatal mortality detection	100-500
Pyelography, intravenous	400-2,000, multiple films and radio-opaque
Pyelography, retrograde	450-1,000, multiple films and radio-opaque
Renal arteriography *	2,000-4,200
Retroperitoneal study	500
Salpingography	10-50
Sella turcica	1-5
Shoulder	Negligible
Sialography	1-10
Sinus series	1-10 (paranasal)
Skeletal series for metastases	50-200
Skeletal maturity detection	Negligible (one hand, one knee, or more)
Skull series	1-10 (av. 4)
Spine, lumbar (sacrum and coccyx)	214-2,000 with lateral and oblique views
Splenoportography	2,000
Small intestine survey	500-1,000 (without fluoroscopy)

---

* Denotes procedures which are high risk to the fetus because they may deliver 4 rads or more to it.

Temporo-mandibular joints	1-10
Urethrocytography (like K.U.B. and cystogram but includes urethra)*	To 600 with films, to 6,000 with fluoroscopy, and to 20,000 with cine
Venocavography	1,000+
Ventriculography	2-20

Major References for Table of Diagnostic Radiography **

Bewley, D. K., Laws, J. W., and Myddleton, C. J., 1957. Maternal and fetal radiation dosage during obstetric radiography examination. *Brit. Journ. Radiology* 30:286-290.

Brent, R. 1960. The effect of irradiation of the mammalian fetus. *Clinical Obstetrics and Gynecology* 3:928-950.

Brent, R. 1972. Irradiation in pregnancy, In *Gynecology and Obstetrics*. Davis, ed. 32:1-30 Harper and Row.

Brown, M. L. 1965. Population dose from x rays. PHS Pub. #2001. ICRP-ICRU, 1969. Exposure of man to ionizing radiation arising from medical procedures with special references to radiation induced disease. An inquiry into methods of evaluation. *Physics in Medicine and Biology* 6:199-258.

Kinlen, L. J. and Acheson, E. D. 1968. Diagnostic irradiation. Congenital malformations and spontaneous abortion. *Brit. Journ. Radiology* 41:648-654.

Michigan Survey. 1961. Medical Radiological Health Data, "Michigan survey of medical radiology experience during pregnancy." 2:159-164.

Peter Bent Brigham Hospital Report. Peter Bent Brigham Hospital, 721 Huntington Ave., Boston, Mass., 1964.

Reekie, D. and Davison, M. 1967. The radiation hazard in radiography of the female abdomen and pelvis. *Brit. Journ. Radiology* 40:849-854.

Schwartz, Gerhard, M.D., Director, Department Radiology, New York Eye and Ear Infirmary, also Fellow American College of Radiology (FACR), Clinical Professor of Radiology, New York Medical College, Practice 33 years. (Consultant on table of exposures to the fetus.) Personal communication, 1972, 1973.

---

* Denotes procedures which are high risk to the fetus because they may deliver 4 rads or more to it.

** This list was obtained from a paper entitled:
"X-ray Effects on the Embryo and Fetus: A Review of Experimental Findings" by Robert Rugh and William Leach, Division of Biological Effects, Bureau of Radiological Health, FDA, United States Public Health Service, Rockville, Maryland 20852 (September 1973).

APPENDIX C: AVERAGE DOSES FOR TYPICAL X-RAY EXAMINATIONS IN MILLIRADS

	Average skin dose per film (a)	Average integral bone marrow dose per exam (b)
Mammography	1500†	—
Upper GI	710	300
Thoracic Spine	980	200
Lower GI	1320	600
Lumbosacral Spine	2180	200
Lumbar Spine (LS)	1920	200
Intravenous Pyelogram (IVP)	590	300
Cervical Spine	240	50†
Cholecystography	620	100
Abdomen or KUB	670	100
Skull	330	50
Lumbo-pelvic	610	100
Chest (radiographic)	44	40
Dental (whole mouth)	910	20
Hip or Upper Femur (thigh)	560	100
Shoulder	260	50†
Dental (bitewing)	920	4
Extremities	100	<10

References

(a) U.S., Department of Health, Education, and Welfare (FDA) Publication 73-8047, *Population Exposure to X rays U.S. 1970*, (Rockville, Md.: Public Health Service, November 1973). Appendix III.

(b) International Commission on Radiological Protection Publication no. 16, *Protection of the Patient in X-ray Diagnosis* (New York: Pergamon Press, 1970).

Average no. of films per exam (a)	Estimated "effective" dose per exam (c)	Average gonadal dose per exam (d)	
		M	F
2*/per breast	300-600	—	—
4.3	150-400	30	150
3*	150-400	<10	<10
2.9	90-250	200	800
3.4	70-250	1000	400
2.9	50-180	1000	400
5.3	50-150	1300	800
3.7	40-80	<10	<10
3.3	25-60	5	150
1.6	10-60	500	500
4	20-50	<10	<10
1.4	5-35	700	250
1.6	5-35	<10	<10
16*	10-30*	<10	<10
3*	2-25*	1200	500
2*	2-25	<10	<10
3*	<5*	<2	<2
2.7	<5*	<10	<10

(c) Preliminary estimates based on work in progress: P. W. Laws and M. Rosenstein, "A Somatic Dose Index for Diagnostic Radiology," to be presented at the Twenty-second Annual Meeting of the Health Physics Society (Atlanta, Ga.: 3-8 July 1977).

† *Ionizing Radiation: Levels and Effects*, 1, *A Report of the United Nations Scientific Committee on the Effects of Atomic Radiation* (New York: United Nations, 1972), pp. 162, 164.

* Estimate by author.

DOSES FOR TYPICAL X-RAY EXAMINATIONS

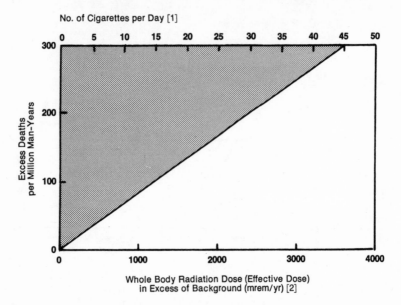

Fig. 1 — Comparison of excess deaths from lung cancer due to cigarette smoking and estimated average excess deaths from leukemia and cancer due to exposure to diagnostic x rays and other ionizing radiation

1. *The Consumers Union Report on Smoking and the Public Interest,* (Consumers Union, Mt. Vernon, N.Y., 1963), p. 34.

2. National Academy of Sciences, National Research Council, *The Effects On Population of Exposure to Low Levels of Ionizing Radiation* (BEIR Report), November, 1972, p. 91.

APPENDIX D: PERSONAL RADIATION
RECORD—DIAGNOSTIC X-RAYS

Name ....

Sex ....

Address ....

Birthdate ....

Date	Physician or Dentist	Location of x-ray facility	Type of Examination	Estimated effective dose*	Estimated gonadal dose*	Details (shielding, mAs, filtration, kVp, etc.)	Purpose of x-ray examination

* See Appendix C for estimate of doses for various procedures conducted under average conditions.

APPENDIX E: FEDERAL AGENCIES AND OTHER ORGANIZATIONS CONCERNED WITH THE USE OF DIAGNOSTIC X-RAYS

A. Federal Agencies Responsible for Radiation Protection:

Bureau of Radiological Health (Department of Health, Education, and Welfare)

> 5600 Fishers Lane
> Rockville, MD 20852

As part of the Food & Drug Administration, BRH is responsible for protecting consumers against unnecessary radiation exposure, particularly from medical sources (including x rays) and from consumer products such as microwave ovens.

National Institute of Occupational Safety & Health (Department of Health, Education, and Welfare)

> 5600 Fishers Lane
> Rockville, MD 20852

This component of HEW is concerned with protecting workers from all on-the-job health hazards including occupational exposure to radiation as might occur from the industrial uses of radioactive materials.

Environmental Protection Agency

> 401 M Street, SW
> Washington, DC 20460

The function of EPA is to set standards which limit contaminants in the general environment, including radioactive pollutants from medical and industrial sources. This agency also regulates the use of medical and dental x rays in federal health care establishments.

B. Other Organizations

American Academy of Dental Radiology
School of Dentistry
University of Florida
Gainesville, FL

American Association of Physicists in Medicine
335 East 45th Street
New York, NY 10017

American Board of Radiology
Kahler East
Rochester, MN 55901

American College of Radiology
20 N. Wacker Drive
Chicago, IL 60606

American Dental Association
211 East Chicago Avenue
Chicago, IL 60611

American Medical Association
535 N. Dearborn Street
Chicago, IL 60610

American Public Health Association
1015 18th Street, NW
Washington, DC 20036

American Radiography Technologists
Bass Building
Enid, OK 73701

American Roentgen Ray Society
Emory University Clinic
Atlanta, GA 30322

American Society of Radiologic Technologists
645 N. Michigan Avenue, Room 620
Chicago, IL 60611

American Veterinary Radiology Society
1201 Waukegan Road
Glenview, IL 60025

Association of University Radiologists
c/o Melvyn H. Schreiber, MD
University of Texas
Medical Branch
Galveston, TX 77550

Health Physics Society
PO Box 156
E. Weymouth, MA 02189

Health Research Group
(Public Citizen)
2000 P Street, NW
Washington, DC 20036

Radiation Research Society
c/o Richard J. Burk, Jr.
4211 39th Street, NW
Washington, DC 20016

Radiological Society of North America
713 East Genesee Street
Syracuse, NY 13210

APPENDIX F: INFORMATION ABOUT BREAST SELF-EXAMINATION AND SCREENING CENTERS

Breast tissue is normally slightly lumpy. Women should examine their own breasts regularly in order to detect changes. If you feel a suspicious lump see a doctor as soon as possible. Many women put this off out of fear. Very few lumps are actually cancerous, but early treatment of cancerous lumps is important. It doesn't pay to wait.

BREAST SELF-EXAMINATION *

Follow these rules:

- Do this examination each month after your menstrual period so that you can be familiar with your breasts in their normal state.
- After menopause, check breasts each month on the first day of a new month.
- See your doctor without delay if any unusual lumps or dimples are noted.

EXAMINATION PROCEDURE: *

(1) Lie down with the right hand under your head. Examine your right breast with your left hand. Push down gently with your fingers flat until you can feel the chest muscle underneath.

(2) In your mind's eye, divide the breast into four areas like a clock: 12:00; 3:00; 6:00 and 9:00. Starting at 12 (at the nipple), move

---

* Reprinted from *How to Be Your Own Doctor* (*Sometimes*) by Keith W. Sehnert, M.D. and Howard Eisenberg, © 1975, by Keith W. Sehnert, M.D. and Howard Eisenberg. Reprinted by permission of Grosset and Dunlap, Inc.

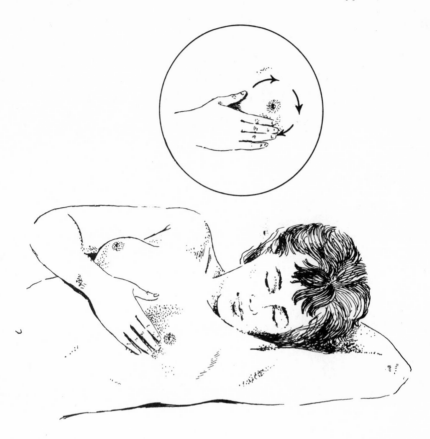

your fingers clockwise from 12 to 3 and on around in a little circle. Then, move your hand up 2 inches and make another circle in the same way; then up another 2 inches etc., until entire breast is covered. (If you notice a lump at, for example, 4 o'clock, you have a point of reference for yourself and your doctor.) Repeat this procedure with your left breast and right hand.

For films and literature on self-examination, write to:

The American Cancer Society

219 E. 42nd Street

New York, NY 10017

(217-867-3700)

Breast Cancer Screening Centers

In early 1975, the American Cancer Society and the National Cancer Institute established 27 breast screening centers around the United States. By making an appointment, women over 35, without symptoms of breast disease, can have a free examination. It will include breast examination (palpation), x-ray (mammography), and heat graph (thermograph). Upon request the mammographic x-ray examination may be omitted for women under 50.

The centers are listed below:

EAST

Rhode Island Hospital
Rhode Island Dept. of Health
Eddy Street
Providence, RI 02908
401-831-6970

Guttman Institute
200 Madison Ave.
New York, NY 10016
212-689-9797

College of Medicine and Dentistry
of N.J.
15 S. 9th St.
Newark, NJ 07107
201-484-9221

University of Pittsburgh
School of Medicine/The Falk Clinic
3601 Fifth Avenue
Pittsburgh, PA 15231
412-624-3336

Temple University-Albert
Einstein Medical Center
York & Taber Rds.
Philadelphia, PA 19141
215-567-0559

Wilmington General Hospital
Chestnut & Broom Sts.
Wilmington, DE 19899
302-428-4815

SOUTH

University of Louisville
School of Medicine
601 S. Floyd St.
Louisville, KY 40402
502-583-2894

St. Vincent's Medical Center
Barrs St. & St. Johns Ave.
Jacksonville, FL 32204
804-389-7751, ext. 8491 or 8492

Vanderbilt University
School of Medicine
Nashville, TN 37322
615-322-2501

Duke University/Medical Center
3040 Erwin Rd.
Durham, NC 27705
919-286-7943 or
919-383-1060

Emory University & Georgia
Baptist Hospital
Atlanta, GA 30322
404-355-4940

Georgetown University
Medical School
3800 Reservoir Rd., N.W
Washington, DC 20007
202-625-2183

MIDWEST

University of Kansas
Medical Center
Rainbow Blvd. at 39th St.
Kansas City, KS 66103
913-342-1338

Medical College of Wisconsin
8700 W. Wisconsin Ave.
Milwaukee, WS 53236
414-257-5200

University of Cincinnati
Medical Center
Eden & Bethesda Aves.
Cincinnati, OH 45229
513-872-5331

University of Michigan
Medical Center
396 W. Washington St.
Ann Arbor, MI 48103
313-763-0056

Iowa Lutheran Hospital
University at Penn
Des Moines, IA 50316
515-283-5678

Cancer Research Center
Business Loop
70th & Garth Ave.
Columbia, MO 65201
314-442-7833

WEST

Oklahoma Medical
Research Foundation
800 N.E. 8th St.
Oklahoma City, OK 73190
405-235-8331, ext. 241

Mountain States Tumor Institute
215 Ave. B
Boise, ID 83702
208-345-3590

Virginia Mason
Medical Center
911 Seneca St.
Seattle, WA 98101
206-624-1144

Pacific Health Research
Institute, Inc.
Alexander Young Bldg.
Suite 545
Hotel & Bishop Sts.
Honolulu, HI 96813
808-524-4337

Samuel Merritt Hospital
Breast Screening Center
384 34th St.
Oakland, CA 94609
415-658-8525

University of Arizona
Arizona Medical Center
Tucson, AZ 85724
602-882-7401 or 7402

Los Angeles County
University of Southern
California Medical Center
Los Angeles, CA 90033
213-226-5019

Good Samaritan Hospital and
Medical Center
1015 N.W. 22nd Ave.
Portland, OR 97210
503-228-8331

St. Joseph's Hospital
1919 LaBranch St.
Houston, TX 77002
713-225-3131, ext. 301

APPENDIX G: DIAGRAMS OF ORGANS EXPOSED TO THE PRIMARY X-RAY BEAM IN COMMON DIAGNOSTIC EXAMINATIONS

The names by which doctors and x-ray technicians refer to various x-ray examinations are often confusing to the lay person. To help clarify matters for prospective patients, I have included in this section illustrations depicting important organs and bony structures which are highlighted or fall within the main beam in some of the most common examinations. The rectangles show the approximate size of the main x-ray beam for each exam. Beam size will probably vary somewhat with the technician involved, and also with the size of the patient being examined. In addition, some technicians adjust the beam to a larger-than-necessary size, in order to avoid mistakes.

SKULL
highlights the brain

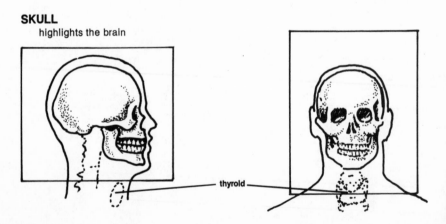

thyroid

THORACIC SPINE
highlights the middle 12 vertebrae of the spine

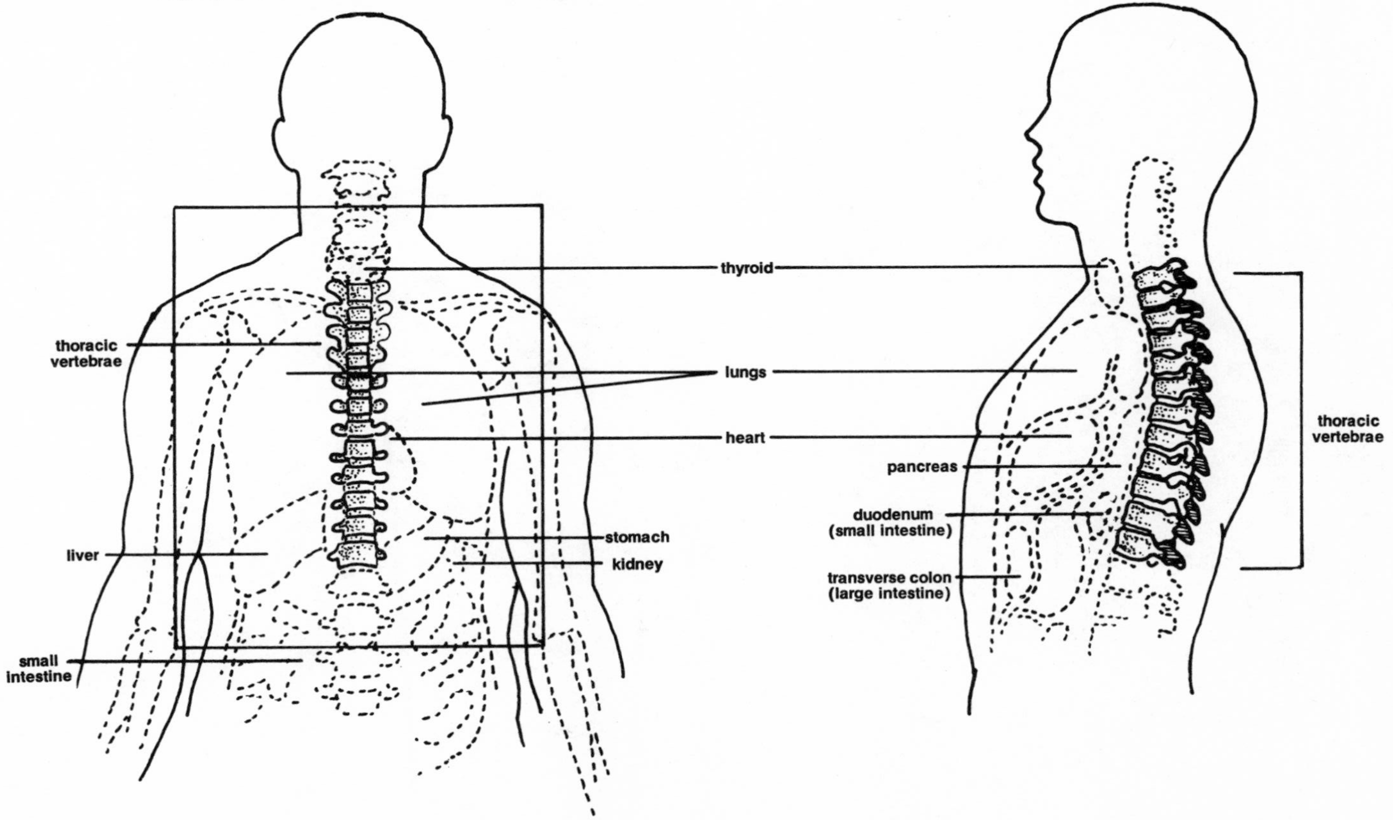

thoracic
vertebrae

thyroid

lungs

heart

pancreas

duodenum
(small intestine)

transverse colon
(large intestine)

stomach

kidney

thoracic
vertebrae

liver

small
intestine

213

CERVICAL SPINE
highlights the upper 7 vertebrae of the spine

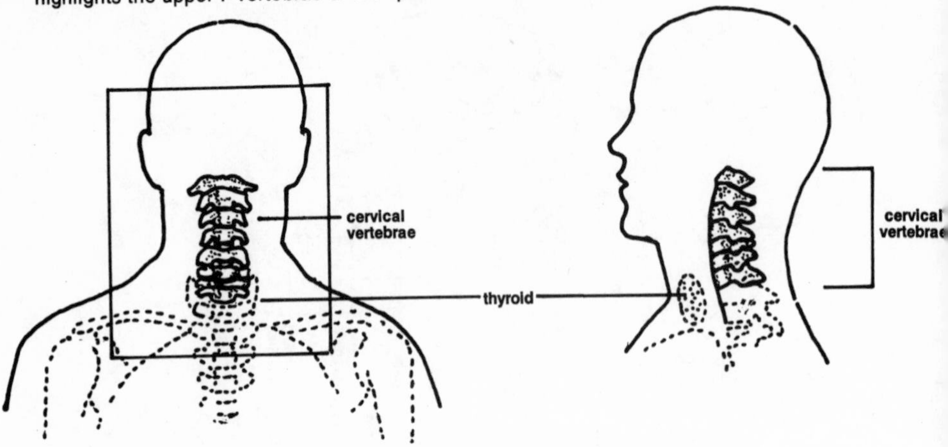

cervical
vertebrae

cervical
vertebrae

thyroid

CHEST
highlights lungs

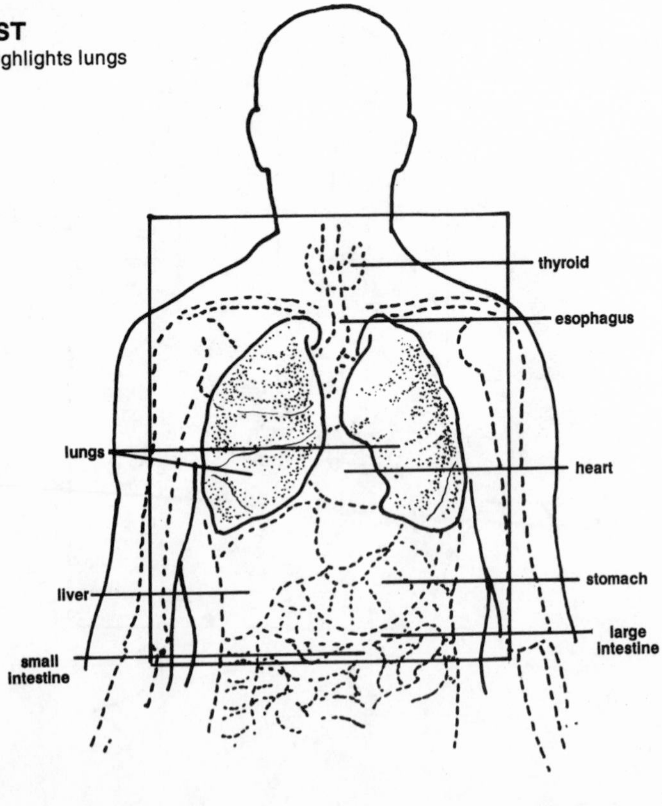

thyroid

esophagus

lungs

heart

liver

stomach

small
intestine

large
intestine

214

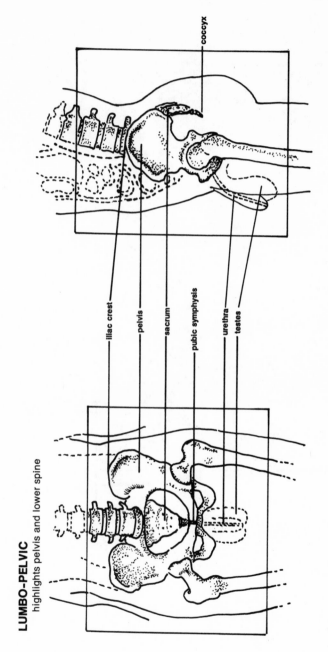

LUMBO-PELVIC
highlights pelvis and lower spine

coccyx

iliac crest

pelvis

sacrum

pubic symphysis

urethra

testes

215

Abdominal exams include the following:
Retrograde pyelogram — highlights kidney, urethra or bladder
Lumbosacral spine — highlights the lower part of the spine, which is
rigid and connected to the pelvis, and the first few sacral vertebrae
KUB — highlights kidneys, ureter and bladder
Intravenous pyelogram (IVP) — highlights kidneys
Renal arteriogram — highlights the blood vessels in and near the kidneys
Barium enema — highlights colon and lower gastrointestinal tract

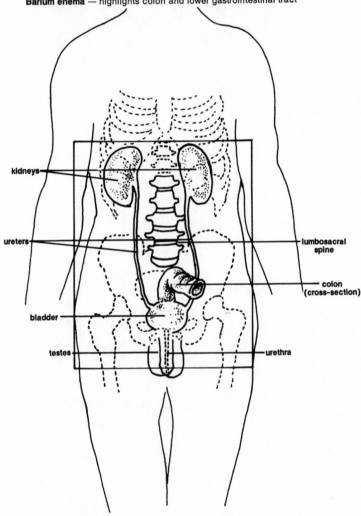

LUMBAR SPINE
highlights lowest 5 vertebrae of the spine

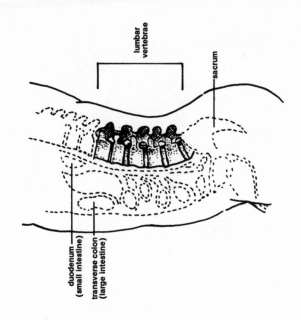

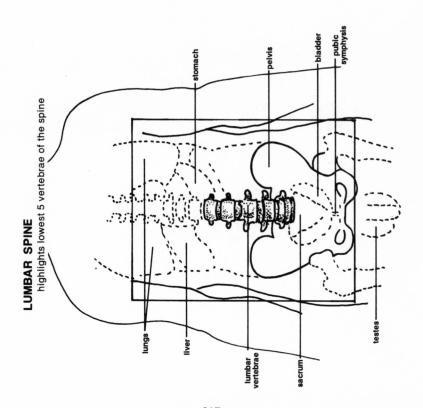

lumbar vertebrae

sacrum

duodenum (small intestine)

transverse colon (large intestine)

stomach

pelvis

bladder

pubic symphysis

lungs

liver

lumbar vertebrae

sacrum

testes

SHOULDER

highlights one shoulder (small rectangle)
or both shoulders (large rectangle)

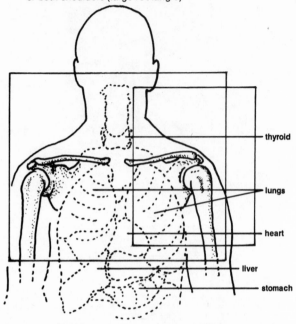

thyroid

lungs

heart

liver

stomach

URETHROGRAM AND CYSTOGRAPHY

highlights urethra and urinary bladder

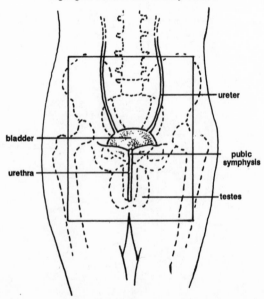

ureter

bladder

pubic
symphysis

urethra

testes

218

RIBS
highlights ribs

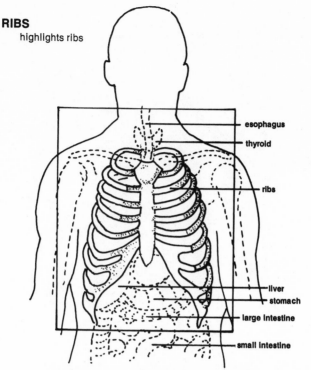

esophagus

thyroid

ribs

liver

stomach

large intestine

small intestine

Note: *the heart and lungs lie behind the rib cage, and are also exposed when the ribs are x-rayed.*

HIP
highlights one hip (small rectangle) or both hips (large rectangle)

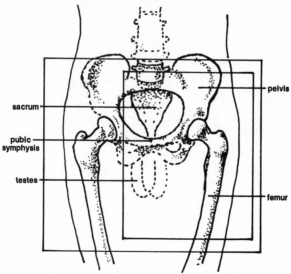

pelvis

sacrum

pubic symphysis

testes

femur

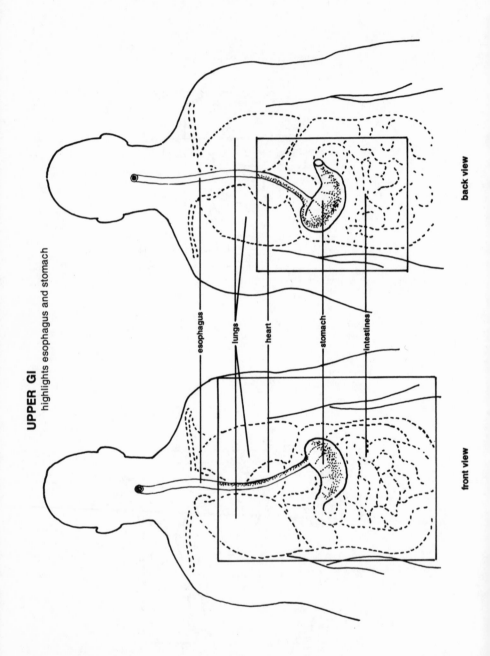

UPPER GI
highlights esophagus and stomach

esophagus

lungs

heart

stomach

intestines

back view

front view

CHOLECYSTOGRAPHY
highlights gallbladder

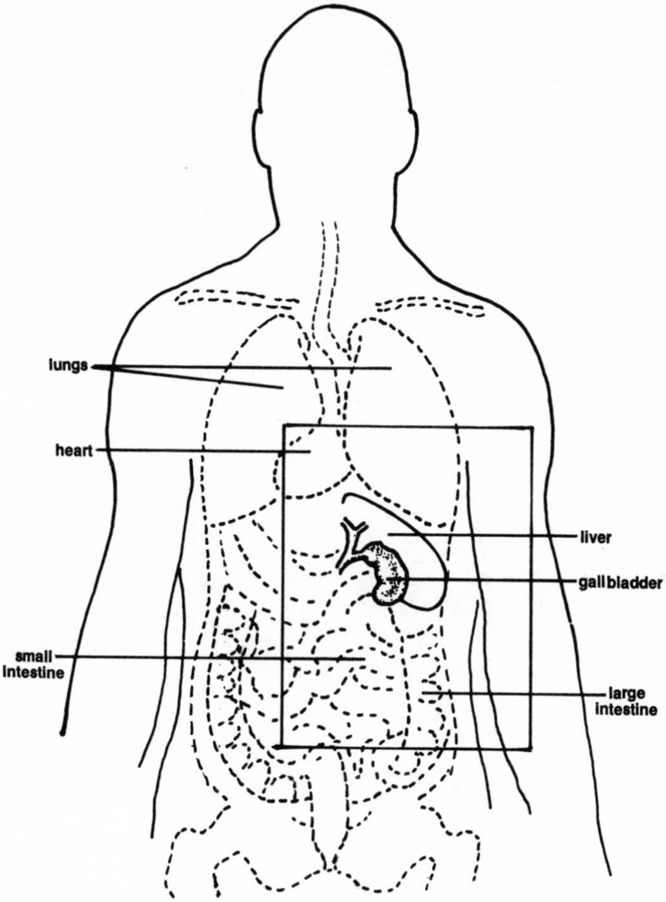

lungs

heart

liver

gallbladder

small intestine

large intestine

back view

HUMERUS OR FEMUR
highlights upper arm or thigh

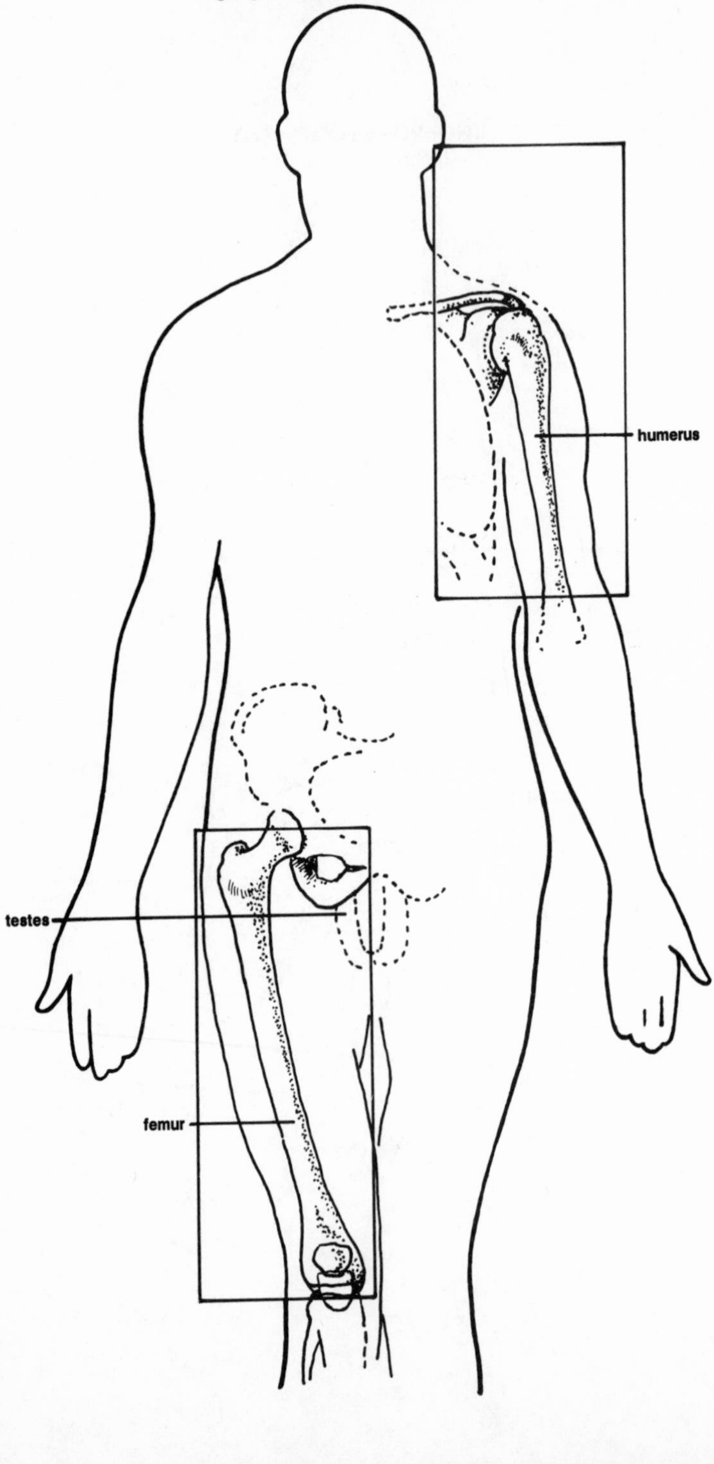

humerus

testes

femur

FULL SPINE
highlights entire spine, and is used by
chiropractors

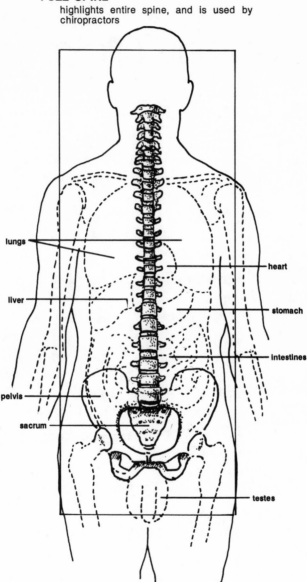

APPENDIX H: GLOSSARY

Terms and abbreviations used in x-ray diagnostics and radiation protection:

Absorbed dose—The energy imparted by x rays or other ionizing radiation to a part of the body (usually tissue or bone) exposed to radiation. The special unit of absorbed dose is the rad which represents 100 ergs of energy absorbed by one gram of material.

Absorption—The process whereby radiation is stopped and reduced in intensity as it passes through matter. Lead, which is denser than most materials, is a good absorber of x rays.

Adaptation (Dark adaptation)—The adaptation of the eye to vision in the dark. The adaptation process is necessary for those operating old-fashioned fluoroscopy units because the image on the fluoroscopic screen is quite dim.

Added filtration—Refers to the filtration (whether aluminum, copper, or lead) placed in the x-ray beam to absorb the less penetrating radiations which do not contribute to the quality of the x-ray image. The use of appropriate filtration prevents unnecessary x-ray exposure.

Air contrast—The introduction of air into a selected part of the body because it does not absorb x rays like the surrounding tissues and hence provides a contrast.

Air contrast study—An x-ray examination performed using air to outline soft-tissue structures within the body. The air may be swallowed, injected, or obtained from carbonated beverages, depending upon the circumstances.

Alpha particle—A form of ionizing radiation consisting of two neutrons and two protons. Alpha particles are often emitted from radioactive materials.

Angiocardiography—X-ray study of the heart and great vessels after the intravenous injection of an opaque fluid.

Angiography—X-ray examination of the blood vessels aided by the injection of a radiopaque contrast substance.

Anode—The anode is the positive electrode within an x-ray tube, toward which electrons are accelerated from the cathode. The kinetic energy possessed by the high-speed electrons is converted to heat and x rays when the electrons strike the anode.

Anterior—Front.

Anterior posterior (A.P.)—An x ray taken from the front to the back.

Aortography—X-ray examination of the aorta after injection of a contrast medium.

Apical (Lordotic)—A term used for a special projection used to obtain a better visualization of the tip of the lungs.

Arteriography—X-ray examination of the arteries aided by the injection of a radiopaque contrast substance.

Atom—A fundamental unit of which matter is composed. It consists of a heavy nucleus and surrounding electrons.

Autoradiography—A radiographic study whereby an x-ray film is exposed by radiations emanating from a radioactive material which is introduced into an organ of the body (or into a plant).

Background—A term used to describe the radiation present in the natural environment. It is produced by radioactive substances in the earth's crust, in water, in air, and by cosmic rays from outer space.

Barium enema (B.E.)—Barium is used as a contrast medium for an x-ray examination of the colon (intestines) or lower gastrointestinal tract.

Barium sulfate—Barium sulfate is a very dense material used in liquid suspension in the barium cocktail or drink used in some x-ray examinations involving the gastrointestinal tract. (*It must not be confused with barium sulfite, which is very toxic.*)

225

Benign—Not malignant. A benign tumor is one whose growth does not spread beyond its original site and is rarely fatal to the host.

Beta particle—A form of ionizing radiation consisting of a fast electron. Beta particles are often emitted from radioactive materials.

Bitewing examination, dental—An x-ray examination of the molars by means of a small paper-covered film which has a small wing-like paper flap that the patient holds in place by biting.

Bone marrow—A soft tissue which constitutes the central filling of many bones and serves as a producer of red blood cells. Bone marrow is especially sensitive to x rays.

Bone survey—An x-ray examination of the bones. This survey is often used to determine the spread of cancer (metastasis).

Bronchography—X-ray examination of the bronchial tree aided by filling the lungs with a radiopaque contrast medium.

Bucky—See Potter-Bucky Diaphragm.

Cancer—A growth (tumor) of abnormal cells which invades the body and may spread through the lymph or blood system (metastasize) causing severe illness or death.

Carcinogenic—Producing cancer.

Carcinoma—Malignant tumors derived from the skin, the membranes lining the body cavities, or certain glands.

Cardboard film holder—A cardboard film holder is used in certain types of x-ray examinations. This type of film holder has no mechanism for intensification of the x rays.

Cardiac catheterization—A procedure whereby a catheter or hollow tube is advanced along a blood vessel into a chamber of the heart for the purpose of obtaining pressure readings, oxygenation determinations, or for the injection of contrast media.

Cassette—A cassette is a kind of x-ray film holder containing intensifying screens mounted within front and back structures which are hinged together. The x-ray film is placed between the intensifying screens in the dark-room to prepare the cassette for use.

Cataract—An opaque body which forms in the eye, obscuring the transparency of the lens. Cataracts can be induced by high doses of radiation.

Cathode—The cathode is the negative electrode in a tube where electrons are produced. It consists of one or two filaments and focusing cups.

Central Beam (Central rays)—Refers to the x rays in the center or most intense part of the x-ray beam.

Cervical spine—X-ray examination of the seven vertebrae located at the top of the spine.

Cholangiography (Post-operative cholangiography)—X-ray examination of the bile ducts following surgery on the gallbladder.

Cholangiography (or **Cholecystography**)—X-ray examination of the gallbladder and/or bile ducts after the intravenous injection of a contrast material which subsequently appears in the bile.

Chromosome—Important rod-shaped molecules found in all body cells. Chromosomes contain the genes of heredity-determining units.

Chronic exposure—Irradiation which is spread out over a period of years. Those who are exposed to radiation on an occupational basis can suffer from chronic exposure.

Collimation—Restriction of the size of the x-ray beam.

Compression cone—This device is an attachment for use in fluoroscopy of the GI tract and serves to permit the examiner to apply pressure to various parts being examined, displace some of the overlying structures, and improve the examination.

Compression device—A compression device is a mechanical means for reducing the thickness of a part of the body for the purpose of improving the x-ray examination. One type is used in IVP work (primarily for compression of the ureters), other types are used in fluoroscopy of the GI tract and mammography.

Cone—A round shield placed in front of the x-ray tube to limit the size of the beam.

Contrast—In radiology, contrast is defined as the difference (in density) between light and dark areas on the processed film. The contrast depends upon 1) applied kilovoltage, 2) filtration, 3) screen and film characteristics, and 4) processing solutions and techniques.

Contrast medium—A substance which, when introduced into a selected part of the body, absorbs x rays differently than surrounding tissues to outline that part of the body more clearly. (See *air contrast* and *radiopaque medium*).

Conventional therapy—X-ray therapy performed with equipment generating voltages up to 250 kilovolts peak (kvp).

Cosmic rays—Very energetic radiation from outer space consisting of x rays and charged particles.

Creep—The horizontal or vertical movement of fluoroscopic equipment during an x-ray examination.

Cystography—X-ray examination of the urinary bladder aided by the introduction of a contrast medium.

Definition (detail)—In roentgenography, definition (or detail) refers to the sharpness of structure lines or contour lines on the processed film. Definition (detail) depends upon 1) size of the focal spot, 2) geometry, 3) filtration, 4) motion, 5) grain size of screens, 6) grain size of film, 7) emulsion thickness of film, 8) processing solutions, and 9) processing techniques.

Depth dose—The radiation dose delivered to a point at a given distance below the skin.

Dermatitis—An inflammation of the skin characterized by irritation, color change, lesions, and itching. This condition can be caused by overexposure to x rays or other forms of ionizing radiation.

Discography—An x-ray examination of one or more intervertebral discs obtained by means of direct injection of a contrast medium.

Dose—A term used interchangeably with "dosage" to express the amount of energy absorbed in a unit of mass in an organ or individual. Dose rate is the dose delivered per unit of time. (See also roentgen, rad, and rem.)

Dosimeter—An instrument which measures radiation dose, as, for example, a pocket dosimeter.

Double-contrast study—X-ray examination in which both a contrast medium and air are used simultaneously (or in succession in certain instances) for the purpose of outlining soft-tissue structures within the body.

Electron—A negatively charged part of an atom or molecule. When electrons strike materials at high energy, x rays are produced.

Embryo—The term used to describe a developing human from conception until the fourth month after conception.

Encephalography—X-ray of the brain after the cavities have been filled with a contrast medium.

Expiration film—A term applied to an x-ray examination obtained while the patient holds his breath after expiring (exhaling) his breath as much as possible.

Exposure—A measure of the number of ions produced in air by radiation. The special unit of exposure is the roentgen and it can be easily measured with an ionization chamber. It should not be confused with dose which is not so easy to measure and represents energy deposited in a certain amount of material.

Femur—Thighbone.

Fetus—The term used to describe a developing human embryo from the fourth month after conception to birth.

Film badge—A light-proof film packet which is used for estimating radiation exposure of personnel who work with x rays or other forms of ionizing radiation.

Fluoroscope—A device in which a fluorescent screen is mounted in front of an x-ray tube so that internal organs may be examined when the x-ray shadow is cast on the screen. The fluorescent screen is coated with a special substance which emits light when exposed to x rays. The fluoroscope is used when an examiner would like to observe a continuous process, and the fluoroscopic image is usually amplified and displayed on a television screen.

Fluoroscopy—The practice of examining through the use of an x-ray fluoroscope. This technique is usually used to detect the motion of organs or materials inside the body.

Gallbladder (G.B.)—A membranous sac attached to the liver in which digestive fluid (bile) is stored.

Gamma rays—Photons or bundles of electromagnetic radiation of high energy which are usually more penetrating than x rays. They are produced during decay of radioactive atoms.

Gastrointestinal series (GI)—Includes an examination of the esophagus, stomach and intestines. An upper GI series involves the ingestion of a barium drink for contrast. A lower GI series involves a barium enema for contrast.

G.B. (Gallbladder)—A membranous sac attached to the liver in which digestive fluid (bile) is stored.

G.B. series (Gallbladder series)—See Cholangiography.

Genes—Parts of chromosomes which determine the inherited traits of the offspring. Genes are contained in the nuclei of cells.

Genetic effects—Mutations or other changes which are produced by irradiation of the genes in a cell which might reproduce.

Gonads—Reproductive organs, or sex glands consisting of male testicles or female ovaries.

Grid—A grid is a device similar to a grating whose purpose in radiology is to absorb scatter radiation which would impair the clarity of the image on the x-ray film.

> **Stationary grid**—A grid which does not move.

> **Potter-Bucky Diaphragm**—A grid which moves between the patient and the cassette during the exposure.

HVL (Half-value layer)—The half-value layer is the thickness of a specified material (usually aluminum, copper, or lead) required to decrease the dosage rate of a beam of x rays to one-half its initial value.

Hard x rays—Hard x rays are x rays of high penetrating power.

Hardness—A relative term to describe the penetrating quality of radiation. The higher the energy of the radiation, the more penetrating (harder) is the radiation.

Head-clamp—A mechanical device attached to the x-ray unit in which the patient's head is held during an x-ray examination in order to reduce motion.

Health physicist—A professional who is especially trained in radiation physics and concerns himself with problems of radiation damage and protection.

Heel effect—The heel effect refers to the unequal intensity of the x-ray beam, the intensity being greatest on the cathode side of the beam and least intense on the anode side of the beam. Some of this variation is reduced by use of lead apertures and shutters which limit the periphery of the primary x-ray beam.

High-voltage x rays—Voltage range from 140 to 250 kilovolts peak (kVp).

Hysterosalpinigography—X-ray examination of the uterus and oviducts aided by the introduction of a radiopaque contrast medium.

Inherent filtration—Refers to the filtration effect of the materials (such as glass and oil) making up the wall of the x-ray tube.

Inspiration film—A term applied to an x-ray examination obtained while the patient inhales and holds his breath.

Integral dose—A calculated dose for a portion of the body, determined by 1) the size of the field, 2) the skin dose, and 3) the depth of tissue at which the dose falls to one-half the skin dose.

Intravenous—An injection into the veins.

I.V.P. (Intravenous pyelogram)—Intravenous pyelography is an x-ray examination of the calices and pelves of the kidneys with opportunity for outlining the ureters and bladder. Substances injected intravenously which are subsequently excreted by the kidneys make this study possible.

Ion—An atom or molecule which has one or more of its surrounding electrons separated from it and therefore carries an electric charge.

Ion chamber—An x-ray measuring device in which gas is ionized in proportion to the quantity of x-ray energy passing through the chamber.

Ion pair—A positively charged atom or molecule (ion) and an electron formed by the action of radiation upon a neutral atom or molecule.

Ionization—The process whereby one or more electrons is removed from a neutral atom by the action of radiation.

Ionizing radiation—Radiation such as x and gamma rays, beta particles, and alpha particles which are capable of producing ions in matter.

Kilovolt—Unit of 1,000 volts which is used to describe the energy of x rays. Most x-ray machines in medical use generate 20 to 150 kilovolt x rays.

Kilovolts peak (kVp)—The maximum voltage impressed on the x-ray tube in kilovolts (1 kilovolt=1,000 volts). The kVp determines the energies or "quality" of the x rays in the beam.

Laminograms (Tomograms, Body-section films)—X-ray films of a selected plane, or level, in the body. Other tissues above or below the selected level are blurred-out by intentional motion of the x-ray equipment while the exposure is being made.

Lateral—X-ray examination taken from the side.

> **Right Lateral Decubitus**—Patient lies on right side.

> **Left Lateral Decubitus**—Patient lies on left side.

Lateral projection—The projection in which the x ray passes through the part of interest in a transverse manner. Usually the side where abnormality is suspected is placed near the film.

> **Right Lateral**—with the right side near the film.

> **Left Lateral**—with the left side near the film.

Leukemia—A blood disease which is characterized by overproduction of white blood cells. It may result from overexposure of the bone marrow to radiation or it may generate spontaneously. It is almost always fatal.

Lienography—X-ray examination of the spleen after the injection of radiopaque contrast medium.

Linear hypothesis—The assumption that a dose-effect curve derived from data in the high dose-rate ranges may be extrapolated through the low dose range to zero, *implying that, theoretically, any amount of radiation will cause some damage.*

Lordotic—See Apical.

Low-voltage x rays—Voltage range up to 140 kVp.

Lower gastrointestinal (lower GI)—X-ray examination of the colon (intestine) aided by the use of a barium enema for contrast.

Lumbar spine—X-ray examination of the lowest five spinal vertebrae (bony segments).

Lumbo-sacral spine—Examination of the lowest part of the spine which is rigid and connected to the pelvis and the first few sacral vertebrae (bony segments).

MA (milliamperage)—The electron current flowing across the x-ray tube in milliamps (1000 milliamps = 1 ampere). It determines the quantity of x rays in the beam per second.

Malignant—Tending to become progressively worse and result in death. As applied to a tumor, cancerous and capable of spreading locally or via the lymphatic or blood system.

Mammography—X-ray examination of the breasts.

MAs (milliampere seconds)—The product of the electron current in milliamps in the x-ray tube and the time in seconds that the current is on. The mAs determines the x-ray beam quantity.

Milliampere (mA)—The thousandth part of an ampere. The mA of the low-voltage filament current as well as the mA of the high-voltage circuit and x-ray tube are measured in radiology on milliammeters provided on the x-ray control panel.

Millirad (mrad)—One-thousandth of a rad.

Millirem (mrem)—One-thousandth of a rem.

Milliroentgen (mr)—One-thousandth of a roentgen.

Molecule—A group of atoms bonded together by electrostatic (chemical) forces.

Multimillion-volt x rays—Voltages higher than about 3 million electron volts.

Mutation—A transformation of the gene which may be induced by radiation and may alter characteristics of the offspring.

Myelographic stop—A myelographic stop is a mechanical device used in conjunction with myelographic examinations in order that the fluoroscopy equipment will not touch or displace the needle in place for the lumbar puncture.

Myelography—X-ray examination of the spinal cord and adjacent structures after a radiopaque contrast medium is injected into the spinal canal.

P.A. (Postero-anterior)—The projection in which the x rays enter the area of interest from the back (posterio) and exit through the front (anterior).

Penumbra—In diagnostic radiology, penumbra refers to haziness at the edge of an image due to the factors listed under *Definition*. In therapeutic radiology, penumbra refers to the radiation of tissues beyond the projected bounds of the primary beam and is due in large part to scatter radiation.

Photofluoroscopic x rays—Process by which x rays are projected on a fluoroscopic screen so that the screen can then be photographed with conventional photographic film.

Photo-timing—This is a method for timing x-ray examinations, the duration of the exposure being controlled by the amount of radiation which reaches a sensitive photo-tube behind the cassette. It provides a means for precisely reproducing densities on roentgenograms.

Photon—A bundle of electromagnetic energy. Each photon carries a fixed amount of energy. An x-ray beam consists of photons having energies in the x-ray region of the electromagnetic spectrum.

Pneumoencephalography—An x-ray examination in which spinal fluid is removed from the brain and replaced with air or another gas. This gas acts as a contrast medium.

Portal (port)—The port ordinarily is considered to be the region of skin through which the x-ray beam *enters* the body in therapeutic radiology. However, there is also an *exit* port where the beam leaves the body.

Potter-Bucky Diaphragm (Bucky) See also GRID—A Bucky is a grid between the patient and the cassette which moves during the exposure and whose purpose is to absorb scatter radiation which would otherwise impair the clarity of the image on the x-ray film.

Projection—A term for the position of a part of the patient with relation to the x-ray film.

Pyelography—X-ray examination of the kidney after the injection of a radiopaque contrast medium.

Quality—A term used to describe the penetrating power of x rays or gamma rays and is related to the energies of the photons in the beam.

Quantity—A term used to describe the number of photons in a beam of x rays or gamma rays.

Rad—A special unit of absorbed dose equal to 100 ergs of energy deposited by ionizing radiation-like x rays in one gram of matter.

Radiation—Energetic sub-atomic particles of electromagnetic waves which move at high speeds.

Radioactive material—A substance that contains atoms whose nuclei have excess energy which is given off in the form of ionizing radiation such as alpha particles, beta particles, gamma rays, or fast neutrons. A number of elements like radium and uranium are naturally radioactive. Others can be generated artificially in processes like fission used in the generation of nuclear power.

Radiographic x rays—Process by which x rays are projected directly on film.

Radiography—See roentgenography for definition.

Radiology—The science of radiant energy and radiant substances; especially that branch of medical science which deals with the use of radiant energy in the diagnosis and treatment of disease.

Radiopaque medium—A material which absorbs x rays and hence casts a shadow on the x-ray film or fluoroscopic screen.

Regenerative process—Replacement of damaged cells by new cells.

Rem—A unit of dose equivalent which is the same as a rad for x rays.

Renal—At or near the kidneys.

Retrograde pyelogram—Examination of the kidney, urethra, or bladder after the injection of a radiopaque contrast medium.

Retroperitoneal pneumography—X-ray examination for the display of the transparent membrane lining the abdominal cavity. In this examination a gas is injected into the abdominal region for contrast.

Roentgen—The amount of radiation which is required to produce ions which carry a charge of .000258 coulombs in a kilogram of air.

Roentgenography—Photography by means of x rays. (Roentgenography and radiography are commonly used interchangeably as essentially synonymous terms.)

Roentgenology—The branch of radiology which deals with the diagnostic and therapeutic use of roentgen rays. (X rays are included, but radioactive substances are excluded in roentgenology.)

Rotating anode—A rotating anode is one used in radiology for the purpose of dissipating the heat produced at the anode, thereby making possible longer exposures and greater tube life.

Sarcoma—Malignant tumors which arise in internal organs, lungs, bones, or connective tissue.

Scattered radiation—Radiation which is scattered in any direction by interaction with objects or within tissue.

-Scopy—suffix: denotes the actual performance of an examination.

Seeds—Seeds are sealed sources of radioactive material used in radiation therapy for direct insertion into tumors. They are ordinarily left in place with no intention of removing them subsequently.

Shielding—Material which is interposed between a radiation source and an irradiated site for the purpose of minimizing the radiation hazard. Shielding is usually made of lead which is dense and absorbs radiation easily. Shielding is often used to protect the reproductive organs, testes, or ovaries, from the x-ray beam during an examination.

Shoulder support—A mechanical device attached to the x-ray unit when necessary in order to support the patient's shoulders and body when the x-ray table is placed in certain positions.

Sialography—X-ray examination of the salivary gland or duct after introduction of a radiopaque contrast medium.

Skin dose—A special instance of tissue dose, referring to the dose immediately on the surface of the skin.

Small bowel series—X-ray examination of the small intestine, using a contrast medium.

Soft-tissue film—A term for an x-ray examination performed with relatively low voltage for the purpose of providing optimal contrast for evaluating soft-tissue structures.

Soft x rays—Soft x rays are x rays of low energy (kVp) and penetrating power.

Somatic—Pertaining to all body tissue other than reproductive cells.

Speed factor—With intensifying screens, the speed factor is defined as the ratio of the exposure required without screens to the exposure required with screens to get the same degree of blackening of x-ray films.

Spot-film—A spot-film is an x-ray exposure, using a cassette, made during the course of a fluoroscopic examination.

Stereo—A term used for an x-ray examination in which two films are obtained in rapid succession but with a pre-determined difference in projection in order that, usually with special equipment, the films permit visualization in three dimensions of an area in question.

Sterility—Inability (either temporary or permanent) to reproduce.

Superficial therapy—A term used in radiation therapy for treatment devised to maximize the effect of treatment on the skin while sparing the underlying tissues.

Target—That part of the metal anode or plate which faces the cathode, and is struck by the beam of electrons. The x rays are produced when the electrons strike the anode.

Target angle—The target angle is the angle away from perpendicular at which the electron stream from the cathode strikes the anode target.

Target-film distance—The distance from the x-ray tube target (anode) to the x-ray film.

Target-skin distance (TSD)—The distance from the x-ray tube target (anode) to the skin of the patient where the x-ray beam enters his body.

Teleroentgenography (Teleoroentgenography)—X-ray studies obtained with the tube at least six feet from the film, with the purpose of striving for parallel rays and minimizing distortion.

Thoracic—In the region of the chest.

Thoracic spine—X-ray examination of the middle 12 spinal vertebrae located above the five lumbar vertebrae and below the seven cervical vertebrae.

Threshold hypothesis—The assumption that no radiation injury occurs below a specific dose level. (See *Linear hypothesis*)

Thyroid—The large gland located in the neck in front of the windpipe. This gland secretes a hormone which regulates body growth and metabolism.

Tissue dose—Used to distinguish between measurements made in air and in the body. Strictly speaking, tissue doses should be given in terms of absorbed dose (i.e. rads) but the roentgen unit is often incorrectly employed.

Tomography—Method for blurring out all parts of an x-rayed body except those lying in one plane.

Total filtration—Refers to the total filtration of the x-ray beam provided by both the inherent filtration and the added filtration.

Trendelenberg position—A position in which the patient lies with his head and upper body lower than his hips.

Tube—Usually refers to the glass tube within the head of the x-ray unit wherein x rays are produced as the result of high-speed electrons striking a metallic target (anode).

Tunnel—A tunnel is a device, either part of a fluoroscopy assembly or part of a chest x-ray cassette holder, whereby a cassette can be protected from the x-ray beam before and after it is used to make a radiograph.

Upper Gastrointestinal (Upper GI)—X-ray examination of the esophagus or stomach aided by the introduction of a radiopaque contrast medium.

Upright film—A term, applied sometimes to x-ray examination of the abdomen, in which the examination was obtained with the patient in an upright (or partially upright) position.

Venography—An x-ray examination of the veins, following injection of a contrast medium.

Ventriculography—X-ray examination of the ventricles of the brain after the injection of a contrast medium.

Volume dose—A calculated dose of radiation in gram-roentgens, based upon the air dose and the grams of tissue irradiated.

Xeroradiography (Xeros: dry)—A form of radiography performed without use of x-ray film or fluorescent screens. A selenium-coated metal surface is substituted for x-ray film, and after exposure to x rays is dusted with calcium carbonate powder, producing an etch-like image. This is a photoelectric process and can be repeated as desired.

X rays—Penetrating photons of electromagnetic radiation having wave lengths shorter than visible light. They are usually produced by bombarding a metallic target with fast electrons in a vacuum. These rays are sometimes called roentgen rays after their discoverer, W. C. Roentgen. In nuclear reactions, it is customary to refer to photons originating in the nucleus as gamma rays, and those originating within an atom but outside its nucleus as x rays.

NOTES

Chapter 1

1. Ivan Illich, *Medical Nemesis* (New York: Random House, 1976), p. 13.
2. A. M. Stewart and G. W. Kneale, "Radiation Dose Effects in Relation to Obstetric X Rays and Childhood Cancers," *Lancet* 1 (1970): 1185–88; R. Gibson et al., "Introduction in the Epidemiology of Leukemia Among Adults," *Journal of the National Cancer Institute* 48 (1972): 301–11.
3. International Commission on Radiological Protection, *The Evaluation of Risks From Radiation,* Publication no. 8 (New York: ICRP, 1966), p. 40.
4. John L. McClenahan, "Wasted X Rays" (Editorial), *Radiology* 96 (August 1970): 453–56; Karl Z. Morgan, "Never Do Harm," *Environment* 13, no. 1 (1971): 28; U.S. Department of Health, Education, and Welfare (FDA) Publication 72–8021, *Excessive Medical Diagnostic Exposure* (Talk delivered at the Third Annual National Conference on Radiation Control by Karl Z. Morgan, May 3, 1971), p. 43.
5. U.S., Environmental Protection Agency (ORP/CSD) Publication 72–1, *Estimates of Ionizing Radiation Doses in the United States 1960–2000,* prepared by A. W. Klement, Jr. et al. (Washington, D.C.: Environmental Protection Agency, August 1972), p. 98.
6. U.S., Department of Health, Education, and Welfare (Code of Federal Regulations) Publication, *Environmental Assessment Report: Performance Standard for Diagnostic X-ray Systems and Their Major Components,* 21 CFR 1020.30–1020.32; and *Related Interpretive Policy Concerning the Assembly and Reassembly of Diagnostic X-ray Components,* 21 CFR 1000.16 (Rockville, Md.: Public Health Service, September 1976), pp. 34–5.
7. U.S., Department of Health, Education, and Welfare (FDA) Publication 73–8047, *Population Exposure to X Rays U.S. 1970* (Rockville, Md.: Public Health Service, November 1973), p. 27.
8. *Environmental Assessment Report,* Appendix B, pp. 1–4.
9. Ibid.
10. Joel Griffiths and Richard Ballantine, *Silent Slaughter* (Chicago: Henry Regnery Co., 1972).
11. "A Nightmare Decision," *Good Housekeeping,* February 1975, p. 16.

Chapter 2

1. Ivan Illich, *Medical Nemesis* (New York: Random House, 1976), p. 34.
2. Jerome Frank, *Persuasion and Healing* (New York: Schocken Books, 1961), p. 67.
3. Ibid., p. 66.
4. Sissela Bok, "The Ethics of Giving Placebos," *Scientific American* 231, no. 5 (November 1974): 17–23; A. Leslie, *The American Journal of Medicine* 16 (June 1954): 854–62; Henry K. Beecher, *The New England Journal of Medicine* 274 (June 1965): 1354–60.
5. Rick Carlson, *The End of Medicine* (New York: Wiley, 1975), p. 19.
6. Victor R. Fuchs, *Who Shall Live?* (New York: Basic Books, 1974), p. 30.

Chapter 3

1. U.S., Department of Health, Education, and Welfare (SSA) Publication 75–11909, *Medical Care Expenditures, Prices and Costs: Background Book* (Washington, D.C.: Department of Health, Education, and Welfare, 1975), p. 14.
2. Selig Greenberg, *The Quality of Mercy* (New York: Atheneum, 1972), p. 151.
3. Garrett Hardin, "The Tragedy of the Commons," *Science* 162 (13 December 1968): 1243–48.
4. *Medical Care Expenditures,* p. 88.
5. Ibid., p. 12.
6. Paul M. Ellwood and Michael E. Herbert, "Health Care: Should Industry Buy It or Sell It?" *Harvard Business Review,* July–August 1973, p. 103.
7. Ibid., p. 102.
8. J. H. Broida et al., "Impact of Membership in an Enrolled, Prepaid Population on Utilization of Health Services in a Group Practice," *New England Journal of Medicine* 292 (April 1975): 780.
9. Victor R. Fuchs, *Who Shall Live?* (New York: Basic Books, 1974), pp. 138–41.

Chapter 4

1. William O. Morris, *Dental Litigation* (Charlottesville, Va.: Michie Co., 1972), pp. 213–14; Otha Linton, Director of Governmental Relations, American College of Radiology, August 1976: personal communication.
2. "Who Really Owns Your Medical Records?" *Consumer Reports* 41, no. 8 (August 1976): 454.
3. Staff of *Prevention* magazine, *The Encyclopedia of Common Diseases* (Emmaus, Pa.: Rodale Press, 1976), p. 140.
4. U.S., Department of Health, Education, and Welfare (FDA) Publication 73–8047, *Population Exposure to X Rays U.S. 1970* (Rockville, Md.: Public Health Service, November 1973), p. 92.
5. CBS News, *60 Minutes* 7, no. 7 (16 February 1975).
6. "Chiropractors—Healers or Quacks?" *Consumer Reports* 40, no. 9 (September 1975): 544.

Chapter 5

1. Rick Carlson, *The End of Medicine* (New York: Wiley, 1975), p. 77.
2. U.S. Public Health Service, "Should Asymptomatic Employees Have a Chest X Ray as Part of a Periodic Physical Examination?" paper prepared for the Division of Federal Employee Health Bureau of Medical Services, Health Services Administration, by Francis P. Barletta (1975).
3. Richard Spark, "The Case Against Regular Physicals," *The New York Times Magazine,* 25 July 1976, pp. 10ff.
4. Ibid., p. 40.
5. Donald Vickery and James Fries, *Take Care of Yourself* (Reading, Mass.: Addison Wesley, 1976), p. 12.
6. "The Annual Rip-Off?" *Time,* 26 July 1976, p. 54.
7. W. Weiss, H. Seidman, and K. Boucot, "The Philadelphia Pulmonary Neoplasm Research Project. Thwarting Factors in Periodic Lung Cancer," *American Review of Respiratory Diseases* 3, no. 30 (March 1975): 389–97.
8. *Journal of the American Dental Association* 90 (January 1975): 171.
9. U.S., Environmental Protection Agency Publication 520 14–76–002, *Recom-*

mendations On Guidance for Diagnostic X-ray Studies in Federal Health Care Facilities (Washington, D.C.: Environmental Protection Agency, June 1976), p. 18.

10. Deborah VanBrunt, *Consumer Perspectives on X Rays* (New York: New York Public Interest Research Group, 15 November 1976).

11. U.S., Department of Health, Education, and Welfare (FDA) Publication 73–8074, *Population Exposure to X Rays U.S. 1970* (Rockville, Md.: Public Health Service, November 1973).

12. *Consumer Perspectives.*

13. Arthur Levin, *Talk Back To Your Doctor* (New York: Doubleday, 1975), pp. 75, 89.

14. *Consumer Perspectives.*

15. Henry LaRocca et al., "Value of Pre-employment Radiographic Assessment of the Lumbar Spine," *Industrial Medicine and Surgery* 39 (June 1970): 253–58; J. T. Redfield, "The Low Back X Ray as a Pre-employment Screen Tool in the Forest Products Industry," *Journal of Occupational Medicine* 13 (May 1971): 219–26; American College of Radiology, *Conference on Low Back X Rays in Pre-employment Physical Examinations,* summary report and proceedings (Tuscon, Ariz.: January 1973): 164–66.

16. Ibid., p. 198.

17. *Consumer Perspectives.*

18. U.S., Department of Health, Education, and Welfare (PHS) Publication no. 1519, *Population Dose From X Rays U.S. 1964* (Washington, D.C.: Public Health Service, 1969), p. 132.

19. U.S., Department of Health, Education, and Welfare (FDA) Publication 76–8026, "Pennsylvania's Experience in Mass Screening," *Proceedings of the Seventh Annual National Conference on Radiation Control,* prepared by T. Gerusky (Rockville, Md.: Public Health Service, 1976), p. 148.

Chapter 6

1. I. MacKenzie, "Breast Cancer Following Multiple Fluoroscopies," *British Journal of Cancer* 19 (1965): 1–8.

2. U.S., Department of Health, Education, and Welfare, Bureau of Radiological Health Publication. "Benefit-Risk in Mammography," *Proceedings of the Eighth National Conference on Radiation Control,* prepared by Richard B. Chiacchierini and Frank E. Lundin (Springfield, Ill.: May 2–7, 1976).

3. Robert Pear, "Are Breast X Rays for Cancer Good or Bad?" *The Washington Star,* 19 March 1976, p. A–3.

4. National Cancer Institute, Correspondence to Project Directors of the Breast Cancer Detection Demonstration Projects, 23 August 1976.

5. "Are Breast X rays for Cancer Good or Bad?"

6. U.S., Department of Health, Education, and Welfare (FDA) Publication 76–8026, "Survey of Mammographic Exposure Levels and Techniques in Eastern Pennsylvania," *Proceedings of the Seventh Annual Conference of Radiation Control Directors,* prepared by Henry Bicehouse (Hyannis, Mass.: April 27–May 2, 1976), pp. 136–43.

7. John C. Bailar, "Mammography: A Contrary View," *Annals of Internal Medicine,* January 1976, pp. 78ff.

8. Deborah VanBrunt, *Consumer Perspectives on X Rays* (New York: New York Public Interest Research Group, 15 November 1976).

9. U.S., Public Health Service (FDA), "X-ray Effects on the Embryo and Fetus: A Review of Experimental Findings," prepared for the Bureau of Radiological

Health by Robert Rugh and William Leach (Rockville, Md.: September 1973);
U.S., Radiological Health Sciences Education Project Publication no. 874, "A Concept and Proposal Concerning the Radiation Exposure of Women," prepared
under contract no. FDA 72–13 with the Bureau of Radiological Health (FDA),
by Reynold F. Brown et al. (Rockville, Md.: Bureau of Radiological Health);
A. M. Stewart and G. W. Kneale, "Radiation Dose Effects in Relation to Obstetric X Rays and Childhood Cancers," *Lancet* 1 (1970) : 1185–88; Irwin D. J.
Bross and Nachimuthu Natarajan, "Leukemia from Low Level Radiation: Identification of Susceptible Children," *New England Journal of Medicine* 287, no. 3
(20 July 1972): 107–10.
10. "A Concept and Proposal," p. 9.
11. U.S. Department of Health, Education, and Welfare, *X-ray Examinations—
A Guide to Good Practice* (Rockville, Md.: Bureau of Radiological Health, 1970),
p. 6.
12. U.S. Department of Health, Education, and Welfare (FDA) Publication 75–
8029, "Considerations of Possible Pregnancy in the Use of Diagnostic X Rays,"
Health Physics in the Healing Arts, Seventh Midyear Topic Symposium, Health
Physics Society, December 1972, p. 599.
13. *Consumer Perspectives.*
14. U.S., Department of Health, Education, and Welfare, Public Health Service
Publication no. 1000, series 22, no. 5, *Medical X-ray Visits and Examinations
During Pregnancy U.S. 1963,* prepared by Morton L. Brown and Arne Nelson
(Washington, D.C.: Department of Health, Education, and Welfare, June 1968),
p. 7.
15. Kevin Kelly et al., "The Utilization and Efficiency of Pelvimetry," *American
Journal of Roentgenology* 125, no. 1 (September 1975): 66–74.
16. U.S., Department of Health, Education, and Welfare, *Environmental Assessment Report for Proposed Guidelines for Use of Specific-Area Gonad Shielding
During Diagnostic X-ray Procedures* (Rockville, Md.: Bureau of Radiological
Health, August 1975), pp. 2, 13.

Chapter 7

1. U.S., Department of Health, Education, and Welfare, "Recent Developments in
Medical Malpractice" (Presentation before the Institute of Medicine Ad Hoc Advisory Committee on Medical Malpractice), prepared for the Office of the Assistant
Secretary for Health by Sharman K. Stephens, March 1976, p. 3.
2. Ibid., p. 2.
3. David S. Rubsamen, "Medical Malpractice," *Scientific American* 235, no. 2
(August 1976): 18.
4. Ibid., p. 20.
5. Ibid., p. 18.
6. Everett J. Gordon, *A Practical Medico-Legal Guide for the Physician* (Springfield, Ill.: Charles C Thomas, Co. 1973), pp. 66–7.
7. Robert Brook et al., "Effectiveness of Non-Emergency Care via the Emergency
Room," *Annals of Internal Medicine* 78 (1973).
8. F. Roberts and C. Shopfner, "Plain Skull Roentgenograms in Children with
Head Trauma," *American Journal of Roentgenology* 114 (1972): 230–34.
9. Nash D. Harwood et al., "The Significance of Skull Fractures in Children. A
Study of 1,187 Patients," *Radiology* 101 (1971): 151–55.
10. R. S. Bell and J. W. Loop, "The Utility and Futility of Radiographic Skull

Examinations for Trauma," *New England Journal of Medicine* 284 (4 February 1971): 236–39.

11. Ibid.

12. U.S., Department of Health, Education, and Welfare Publication no. 73–88, *Report of the Secretary's Commission on Medical Malpractice* (16 January 1973), p. 13.

13. "The Medical Malpractice Threat: A Study of Defensive Medicine," *Duke University Law Review,* 1971, pp. 939–93.

14. "Defensive Medicine and the Radiology Department," *Practical Radiology,* 9 March 1973.

15. *Medical Opinion,* December 1971.

16. R. H. Blum, "The Psychology of Malpractice Suits," California Medical Association (San Francisco: 1957).

17. *Report of the Secretary's Commission,* p. 69.

18. Deborah VanBrunt, *Consumer Perspectives on X Rays* (New York: New York Public Interest Research Group, 15 November 1976).

19. *Does the Malpractice Insurance Crisis Affect You?* Pennsylvania Medical Society Pamphlet (20 Erford Road, Lemoyne, Pa. 17043).

20. *A Practical Medico-Legal Guide,* p. 66.

21. *Does the Malpractice Insurance Crisis Affect You?*

22. *Report of the Secretary's Commission,* p. 18.

23. Ibid., p. 19.

24. Ibid., p. 10.

25. Ibid.

26. Ibid., p. 12.

27. "Medical Malpractice," p. 18.

28. William O. Morris, *Dental Litigation* (Charlottesville, Va.: Michie Co., 1972), p. 86; Angela R. Holder, "Non-Negligent Failure to Take X-ray Films," *Journal of the American Medical Association* 219 (28 February 1972): 1259–60; Smith v. Yohe (194 A.2d 167 p. 171); Roll v. Ferry (235 F.Supp. 821); Gresham v. Ford (241 SW 2d 408); Carrigan v. Roman Catholic Bishop (178 A2d 502); See also: 162 A.L.R. 1295N. and cases collected therein.

29. "Non-Negligent Failure."

30. Unpublished data from the Bureau of Radiological Health, prepared by James Morrison of the Division of Training and Medical Application (Rockville, Md.: Bureau of Radiological Health, June 1976).

Chapter 8

1. James E. Harris and Kent R. Weeks, *X-raying The Pharaohs* (New York: Scribners, 1973), pp. 22ff.

2. "An X-ray Machine for Babies," *Parade,* 5 January 1975.

3. Arthur L. Robinson, "Image Reconstruction (I): Computerized X-ray Scanners," *Science* 190 (7 November 1975): 542.

4. Ibid., p. 543.

5. Quotation from Jonathan Spivak, "A 'Glamor Machine' Is Hailed by Doctors as a Boon to Diagnosis," *Wall Street Journal* 186, no. 114 (10 December 1975).

6. Massachusetts Department of Public Health, Report on CT Scanning (24 February 1976, attachment IVk; "A 'Glamor Machine' Is Hailed."

7. "A 'Glamor Machine' Is Hailed."

8. U.S., Department of Health, Education, and Welfare. *Bulletin of the Bureau*

of Radiological Health, Supplement no. 1 (Rockville, Md.: Department of Health, Education, and Welfare, July 1976), pp. 2–9.

9. U.S., Department of Health, Education, and Welfare (FDA) Publication 73–8009, "The Medical Problem of Dose Reduction," *Proceedings of a Symposium on the Reduction of Radiation Dose in Diagnostic X-ray Procedures,* prepared by Richard Chamberlain, (Houston, Texas), pp. 100ff.

10. E. F. Schumacher, *Small Is Beautiful* (New York: Harper and Row, 1973), p. 154.

11. Frederick Ferré, *Shaping The Future* (New York: Harper and Row, 1976), pp. 41ff.

12. Ivan Illich, *Tools for Conviviality* (New York: Harper and Row, 1973), pp. 10ff.

Chapter 9

1. Jacques Ellul, *The Technological Society,* trans. J. Wilkinson (New York: Vintage Books, 1964), p. 21.

2. U.S., Department of Health, Education, and Welfare (FDA) Publication 76–8016, *A Study of Retakes in Radiology Departments of Two Large Hospitals,* prepared by Bruce M. Burnett et al. (Rockville, Md.: Department of Health, Education, and Welfare, July 1975), and the references cited therein.

3. Interview of Benjamin Felson, M.D., *Medical World News,* February 1975.

4. *A Study of Retakes; D. Bourne,* "Repeats—An Aspect of Department Management," *Radiography* 35 (1969): 257–61; I. P. Leggett, Jr., W. W. Schadt, and L. C. MacDonnell, "X-ray Film Retakes Rates at Selected Hospitals in the District of Columbia," report by the District of Columbia Department of Human Resource, Health Services Administration, Bureau of Public Health Engineering, Radiological Health Division (February 1971); P. Garner, "Oh No!! Not Again," *Ark Sparks, The Official Publication of the Arkansas Society of Radiologic Technologists* 20, no. 3 (Fall 1970); R. H. Morgan and J. C. Gehret, "The Radiant Energy Received by Patients in Diagnostic X-ray Practice," *American Journal of Roentgenology, Radium Therapy and Nuclear Medicine* 97 (1966): 793–810.

5. U.S., Department of Health, Education, and Welfare (FDA) Publication 74–8028, *Gonadal Shielding in Diagnostic Radiology* (Rockville, Md.: Bureau of Radiological Health, June 1974), p. 3.

6. U.S., Department of Health, Education, and Welfare (FDA) Publication 73–8047, *Population Exposure to X Rays U.S. 1970* (Rockville, Md.: Public Health Service, November 1973), p. 100.

7. M. Neuweg and P. Brunner, "Radiation Exposure Limits in the Healing Arts," *Applied Radiology,* November–December 1974, p. 35.

8. Background information for the DTMA Subcommittee of the Bureau of Radiological Health Medical Radiation Advisory Committee on Proposed Project to Develop Suggested National Standards for Qualification of Medical Radiation Technologists (Rockville, Md.: Bureau of Radiological Health, 27–28 January 1976).

9. Ibid.

Chapter 10

1. Morton Weisinger, "How the Wealthy Stay Healthy," *Parade,* 22 August 1976, p. 14.

2. Ibid., p. 16.

3. "Order More Diagnostic Tests? Making Your Practice More Malpractice

245

Proof," *Medical Economics,* 30 September 1974, p. 77.

4. Donald Vickery and James Fries, *Take Care of Yourself* (Reading, Mass.: Addison Wesley, 1976), p. 5.

5. Ibid., p. 6.

6. Selig Greenberg, *The Quality of Mercy* (New York: Atheneum, 1971), p. 155.

Chapter 11

1. U.S., Department of Health, Education, and Welfare (FDA) Publication 73–8047, *Population Exposure to X rays U.S. 1970* (Rockville, Md.: Public Health Service, November 1973), p. 54.

2. Jean L. Marx, "Diagnostic Medicine: The Coming Ultrasonic Boom," *Science* 186 (18 October 1974): 247–50.

Chapter 12

1. U.S., Environmental Protection Agency (ORP/SID) Publication 72–1, *National Radiation Exposure in the United States,* prepared by D. T. Oakley (Washington, D.C.: Environmental Protection Agency, June 1972), p. 42.

2. National Academy of Sciences, National Research Council, *The Effects on Population of Exposure to Low Levels of Ionizing Radiation* (BEIR Report), November 1972, p. 42.

3. International Commission on Radiological Protection Publication no. 8, *The Evaluation of Risks from Radiation* (New York: ICRP, 1966); *Ionizing Radiation Levels and Effects,* 1 and 2. *A Report of the United Nations Scientific Committee on the Effects of Atomic Radiation* (New York: United Nations, 1972).

4. R. Seltser and P. Sartwell, "The Influence of Occupational Exposure to Radiation on the Mortality of American Radiologists and Other Medical Specialists," *American Journal of Epidemiology* 81 (1965): 2–22; U.S., Bureau of Radiological Health, "Current Mortality Experience of Radiologists and Other Physician Specialists," prepared by G. M. Matanoski for the Division of Biologic Effects Seminar, 12 December 1973.

5. *The Evaluation of Risks,* pp. 4ff.

6. Ibid., p. 4.

7. W. M. Court-Brown and R. Doll, "Mortality from Cancer and Other Causes after Radiotherapy for Ankylosing Spondylitis," *British Medical Journal* 5474 (1965): 1327–32.

8. *The Effects on Population,* p. 68.

9. *The Evaluation of Risks,* p. 40.

10. A. M. Stewart and G. W. Kneale, "Radiation Dose Effects in Relation to Obstetric X Rays and Childhood Cancers," *Lancet* 1 (1970): 1185–88.

11. Irwin D. J. Bross and Nachimuthu Natarajan, "Leukemia from Low Level Radiation: Identification of Susceptible Children," *New England Journal of Medicine* 287 (1972): 107–10.

12. R. Gibson et al., "Irradiation in the Epidemiology of Leukemia Among Adults," *Journal of the National Cancer Institute* 48 (1972): 301–11.

13. U.S., Public Health Service (FDA), "X-ray Effects on the Embryo and Fetus: A Review of Experimental Findings," prepared by Robert Rugh and William Leach for the Bureau of Radiological Health (Rockville, Md.: Bureau of Radiological Health, September 1973), p. 2.

14. International Commission on Radiological Protection Publication 14, *Radiosensitivity and Spatial Distribution of Dose* (New York: ICRP, 1966), pp. 37ff.

15. Ibid., pp. 19–20.

16. *The Effects on Population,* chapter 5.
17. U.S. Department of Health, Education, and Welfare, *X-ray Examination—A Guide to Good Practice* (Rockville, Md.: Bureau of Radiological Health, 1971), p. 17.

Chapter 13

1. *The Consumers Union Report on Smoking and the Public Interest* (Mt. Vernon, N.Y.: Consumers Union, 1963), p. 34.
2. International Commission on Radiological Protection Publication 14, *Radiosensitivity and Spatial Distribution of Dose* (New York: ICRP, 1966), appendix 4, pp. 74ff.; National Academy of Sciences, National Research Council, *The Effects on Population of Exposure to Low Levels of Ionizing Radiation* (BEIR Report), November 1972, pp. 100ff.; *Ionizing Radiation Levels and Effects,* 1 and 2, *A Report of the United Nations Scientific Committee on the Effects of Atomic Radiation* (New York: United Nations, 1972), pp. 402ff.
3. W. Jacobi, "The Concept of Effective Dose Proposed for the Combination of Organ Doses," *Radiation and Environmental Biophysics* 12 (1975): 101–09; P. W. Laws and M. Rosenstein, "A Somatic Dose Index for Diagnostic Radiology," to be presented at the Twenty-Second Annual Meeting of the Health Physics Society (Atlanta, Ga.: 3–8 July 1977).
4. *The Effects on Population,* p. 91.
5. U.S., Environmental Protection Agency (ORP/SID) Publication 72–1, *National Radiation Exposure in the United States,* prepared by D. T. Oakley (Washington, D.C.: Environmental Protection Agency, June 1972), p. 42.
6. Arthur Levin, *Talk Back To Your Doctor* (New York: Doubleday, 1975).

Chapter 14

1. U.S., Department of Health, Education, and Welfare, *X-ray Examinations—A Guide to Good Practice* (Rockville, Md.: Bureau of Radiological Health, 1971), p. 2; U.S., Environmental Protection Agency Publication 520 14–76–002, *Recommendations on Guidance for Diagnostic X-ray Studies in Federal Health Care Facilities,* June 1976, p. 18.
2. U.S., Department of Health, Education, and Welfare (FDA) Publication 73–8047, *Population Exposure to X Rays U.S. 1964* (Rockville, Md.: Public Health Service, 1968), p. 148.
3. U.S., Department of Health, Education, and Welfare (FDA) Publication 73–8047, *Population Exposure to X Rays U.S. 1970* (Rockville, Md.: Public Health Service, November 1973), p. 98.
4. National Academy of Sciences, National Research Council, *The Effects on Population of Exposure to Low Levels of Ionizing Radiation* (BEIR Report), November 1972, p. 13.
5. Over the past several years a number of organizations have recommended the use of shielding for the reproductive organs. These organizations include the American College of Radiology (ACR), the National Council on Radiation Protection (NCRP), The International Commission on Radiologic Protection (ICRP), The National Academy of Sciences (NAS), The Bureau of Radiological Health (BRH), and the Environmental Protection Agency (EPA).
6. U.S., Department of Health, Education, and Welfare (FDA) Publication 74–8028, *Gonadal Shielding in Diagnostic Radiology* (Rockville, Md.: Bureau of Radiological Health, June 1974), p. 5.

INDEX

Numbers in italics refer to diagrams in the appendices.

Beth Israel Hospital, 96-97
Bitewing x rays. *See* Dental x rays
Black lung disease, 60, 112
Bladder, *216, 218*
Blood
    clots, 103
    vessels, 6, 104
Blue Cross, 20-21
Blue Shield, 21
Bone
    absorption of x rays, 135
    broken, 6
    demineralization, 6
    density, 176, 196
    disease, detection of, 104
    disease, in pregnancy, 168
    radiation damage, 157, 185, 191
    x rays and, 8, 40, 43, 59, 102,
        140, 199
Books, about medicine, 125-26
Boston, 96
Bowels, 76
Brain
    abnormalities, 103
    damage, to fetus, 71
    hemorrhage, 103
    problems, diagnosis, 9, 140
    x rays. *See* Skull x rays
Breast cancer, 8, 192-93
    detection clinics, 141
    diagnosis of, 6-7, 48-49, 63-70,
        104, 169, 192-93. *See also* Mam-
        mography; Screening pro-
        grams, breast cancer
        screening centers, 210-11
Breast self-examination, 208-9, *209*
Bronchogram, 166
Brook, Dr. Robert, 88
Bureau of Radiological Health, 60,
    66, 75, 105
    description of, 206
    survey, 113-14

California, 60, 82, 86, 91, 104, 115
Cancer
    breast. *See* Breast cancer
    cervical, 16, 49
    diagnosis of, 6-8, 48-49, 83, 104

increase in, 2
lung, 2, 6, 11, 49, 156, 158, 204
    potential causes of, 66-69, 140,
        150, 157-58
    skin, 8, 55, 156, 184
    and smoking, 156
    thyroid, 8, 184
    treatment of, 16-17, 48
    and radiation, 8, 147, 150, 157-58
    x-ray therapy, 192
    See also Leukemia
Carcinoma, of cervix, 198
Cardiac examinations, 198. *See also*
    Heart
Carlson, Rick, 16, 48, 101
Cataracts, and radiation, 146, 151-52,
    184
Catheterization, 136
CAT-Scanner, 101-6, 137
    cost of, 104-5
Cells, 4
    radiation damage to, 148-49, 152,
        190
    reproduction, 167, 190
Cephalopelvimetry, 198
Cesarean section, 76
Chamberlain, Richard, 106
Checkups
    dental, ix-x, 47-48, 50-53, 168,
        170-71
    medical, 47-48, 59, 122, 163, 168
    See also Employment x rays
Chemicals, and cell damage, 190
Chest, examination, 166
Chest x rays, 48-49, 131, 169-70,
    202-3
    and breast cancer, 192-93
    employees, 170
    for maternity patients, 77-78
    and pregnancy, 198
    risk in, 8
    See also Hospitals, chest x rays
        in; Mobile unit x rays
Childbirth. *See* Pregnancy
Children
    behavior during x rays, 177
    effects of x rays on, 129, 167,
        173, 177

250

X Rays: More Harm Than Good?

Intestines, 199-200, *216-17, 219-21*
  See also Gastrointestinal tract
  x rays
Intravenous Pyelogram (IVP), 166,
  168, 177, 200, 202-3, *216*
Ion pair, 195-96
Japan, radiation victims, 146, 151
Johnson, Dr. G. Timothy, 29, 47
The Journal of Clinical Chiropractic,
  43
Journal of the National Cancer Institute, 66

Kaiser Plan (California), 21, 26
Keim, Dr. Hugo, 42
Kelly, Dr. Kevin, 75-76
Kennedy, Sen. Edward, 1
Kidneys, 6
  examination of, 59, 166, 168,
  199, *216*
  infection, diagnosis of, 6

Lawsuits, 170. *See also* Malpractice,
  suits
Lead
  cylinders, 174, 180
  shields, 179-80. *See also* Shielding
Learning, about medicine, 17, 125-27,
  131
Legal system, 94-95, 97-98
Legislation
  of patient exposure, 114, 131
  of x-ray licensing, 115, 131
Leukemia
  in adult males, 152
  animal studies, 147
  atomic bomb victims, 146
  and children, 150, 191
  and radiation, 4, 7-8, 59, 156-57,
  204
  risk of, 71, 184, 187
Levin, Dr. Arthur, 54, 63, 121, 126
Licensing. *See* X rays, facilities, licensing
Linear hypothesis, 151-52, 157, 197
Literature, medical, 125-27
Liver, *217, 219, 221*
Loop, J. W., 90-91
Lumbar spine. *See* Spine

Lumps. *See* Breast cancer; Tumors
Lungs, *217,* 219, *220-21. See also*
  Cancer, lung
Lymph nodes, 67. *See also* Breast
  cancer

Malpractice
  insurance, 81-82, 91, 94-95
  suits, 13, 29, 92-98
  x rays in, xii, 8, 37, 81-83
Mammography, 63-70, 141, 192-93,
  210
  equipment, 173-74
  radiation dose, 166, 202-3
  as screening tool, 48-49, 169
Managing x-ray technology, 109-19.
  See also Technique, with x rays
Marrow. *See* Bone
Maryland, 38, 88
Mass media, and public attitudes,
  94-95
Mastoid areas, 199.
Maternity patients, 77-78. *See also*
  Pregnancy
Medicaid, 20, 24-25, 44
Medical College of Pennsylvania, 49
Medical costs. *See* Costs
Medical Economics, 122
Medical Nemesis (Illich), 2, 19
Medical profession. *See* Attitudes;
  Doctor-patient relationship; Ridicule, of patients
Medicare, xii, 20, 24-25, 44
Menopause, 66
*Merck Manual of Diagnosis and
  Therapy,* 125
Milk contamination, 77, 184-85. *See
  also* Pregnancy
Millirads, 114, 145
Miscarriage, 71, 188
Mobile unit x rays, 48, 60-61, 131
  and amount of exposure, 169,
  171
  and preemployment tests, 58
Molecules, 141-42, 190, 195-96
Mortality rates, 1-2, 12
Multiple sclerosis, 17
Munich, Germany, 102
Myelogram, 3, 177, 199

254

Nader, Ralph, x
Nagasaki, 146
National Academy of Sciences (NAS), 152, 157
National Breast Cancer Detection Demonstration project, 64
National Cancer Institute, 64, 66, 68, 150, 169
National Council on Radiation Protection and Measurement (NCRP), 152
National Evaluation of X-ray Trends (NEXT), 114-15
National Health Insurance Plan, 25-26
National Institute of Occupational Safety and Health, 206
Nephrotomography, 199
Nerve damage, risk of, 158
Neurologists, 2, 40
New Jersey, 82, 115
New York City, 51, 54, 56, 59, 115
New York Public Interest Research Group, 7, 54, 56, 59, 68, 72, 93
New York state, 59
Non-radiation risk, 158-59
North Carolina, 91
Nuclear scans and scanning, 139, 141-42, 185

Obstetric examinations, 199. *See also* Pregnancy
Obstetricians, 74. *See also* Pregnancy
Ohio, 60
Operators of x-ray equipment, xi, 7-9, 106, 114-19
  awareness of pregnancy, 72, 117
  evaluating, 165, 171, 176-77
  in hospitals, 56
  licensing of, 131
  personal responsibility, 115-19
  protection of, 191
  techniques, 67, 98, 110-15, 139, 165, 212
  training of, 36, 106, 110-11, 114, 118
  See also Radiologists
Ophthalmologists, 2-3

Oral surgeons. *See* Dentists
Organizations, list of, 207
Orthodontists. *See* Dentists
Orthopedic surgeons, 40
Ovaries, 63, 76-77, 113, 176
  and back x rays, 40, 187
  radiation doses, 179
  See also Reproductive organs; Shielding
Ownership, of x rays, 37-39, 188

Pap smear, 48-49
Parade magazine, 122
Patients. *See* Consumers; Doctor-patient relationships; Hospitals
Peer review, 96. *See also* Malpractice
Pelvimetry, 48, 74-76, 140, 187, 198, 200
Pelvis, 187, 215-17
  examination of, 166, 168, 177, 199-200
  x ray of, 39-40, 202-3
Penal institutions, x rays in, 6
Pennsylvania, 60, 67, 82
Pennsylvania Dental Association, ix
Pennsylvania Medical Society, 94
Periodontists. *See* Dentists
Petrous pyramids, 200
Photofluorographic-type x rays, 58
Photofluoroscopic techniques. *See* Fluoroscopy
Physical examinations. *See* Checkups, medical; Employment x rays
Physicians. *See* Medical profession; name of medical specialty
Placebos, x rays as, 14-17
Placenta praevia, 200
Placentography, 200
Pneumoencephalogram, 103, 167, 200
Pneumonia, 6
Polio, prevention of, 12, 16
Polyps, 6
Polytomography, 200
Potchen, E. J., 81
Potter-Bucky diaphragm, 175
A Practical Medico-Legal Guide for the Physician, 84, 94
Prebirth x rays. *See* Fetus; Pregnancy

Preemployment x rays. *See* Employment x rays
Pregnancy
  abdominal examinations in, 7-8, 72, 131, 141, 168
  diagnosis of problems in, 140
  diagnostic procedures, 2, 198-201
  Hiroshima survivors, 146-47
  RH incompatibility tests in, 49
  x ray risks in, 63-64, 70-76, 150, 155, 167-68, 177, 181, 187
  See also Pelvimetry
Prenatal mortality detection, 200
Prenatal x rays. *See* Pregnancy
Prevention magazine, 16
Preventive dentistry, 12
Preventive medicine, 122-24. *See also* Checkups; Health, maintenance of
Protection. *See* Agencies, federal; Operators of x-ray equipment, protection of; Shielding
Psychological use of x rays. *See* Placebos, x rays as
Public Citizen Health Research Group, x-xi, xii, 96, 126, 207
Puerto Rico, 115
Pyelogram, retrograde, *216*

The Quality of Mercy, 130

Rad, 103, 157, 173-74, 195-97
  definition, 145
Radiation, 135
  from atomic bomb, 146-47
  damage to children. *See* Children
  dosage, 4, 7-9, 65, 67, 103, 146-52, 156-58, 163, 166, 173-74, 184, 186, 195-204. *See also* Effective doses
  effects of, ix, 145-52, 155-58, 195-97
  in environment, 2
  exposure to, ix-xi, 4, 39, 60, 67, 103, 105-6, 114, 119, 138-39, 148, 163, 170
  levels of, 8, 152
  risks, 2, 7-9, 142, 156, 166, 185
  from sun, 8, 183-84

  therapy, 24
  ultraviolet, 8, 183-84
Radiation Control for Health and Safety Act, 105, 173
Radioactive
  fallout, 184-85
  substances, 141-42, 184-85
Radiological Health Agency, state, 171
Radiologists, 7, 11, 37, 44, 106, 117-19, 127, 170-71
  evaluating, 163, 165
  rate of cancer, 146
  role of, 110
  use of x rays, 11-12, 67, 70, 98, 103
  See also Operators of x ray equipment
Radiopaque materials, 136, 199
Reader's Guide to Periodical Literature, 126
Rem, 145-46, 149-50, 157, 195
  definition, 145, 195
Renal arteriography, 200, *216*
Reproductive organs, 9, 43-44, 59, 155, 158, 176-77, *217*
  in children, 167
  in dental x rays, 174, 186
  radiation damage, 147-48, 170
  See also Shielding
Respiratory system, disease, 104. *See also* Lungs
Responsibility
  legal, 98. *See also* Malpractice, suits; Ownership, of x-rays
  patient. *See* Attitudes; Health care, personal
  for x rays, 115-19
Retroperitoneal study, 200
Ribs, *219*
Ridicule, of patients, 118, 127-29
Roentgen, 195-97
  definition, 145
Roentgen, Wilhelm, 103, 196
Rubsamen, Dr. David, 82, 95

Sacrum. *See* Spine
Safety requirements, for equipment, 105-6